Global Atmospheric Change
and Public Health

Global Atmospheric Change and Public Health

Proceedings of a Conference Sponsored by
Center for Environmental Information, Inc., 99 Court Street
Rochester, New York 14604-1824

Edited by
James C. White, Ph.D.
Professor Emeritus, Cornell University

Associate Editors
William Wagner
Center for Environmental Information

Carole N. Beal
Center for Environmental Information

Elsevier
New York • Amsterdam • London

Elsevier Science Publishing Co., Inc.
655 Avenue of the Americas, New York, New York 10010

Sole distributors outside the United States and Canada:
Elsevier Applied Science Publishers Ltd.
Crown House, Linton Road, Barking, Essex IG11 8JU, England

© 1990 by Elsevier Science Publishing Co., Inc.

Softcover reprint of the hardcover 1st edition 1990

This book was printed on acid-free paper.

This book has been registered with the Copyright Clearance Center, Inc. For further information please contact the Copyright Clearance Center, Inc., Salem, Massachusetts.

ISBN-13: 978-94-010-6683-9   e-ISBN-13: 978-94-009-0443-9
DOI:10.1007/ 978-94-009-0443-9

Current printing (last digit):
10 9 8 7 6 5 4 3 2 1

Manufactured in the United States of America

CONTENTS

PREFACE

The world is just beginning to face up to the problems which will be brought about by global climate change. Most people equate climate change with rising temperatures, disturbed weather patterns, agricultural crises, and sea level rises; yet potential health effects may be the most significant factors in the whole developing picture.

Man's effect on climate accelerates as population increases. Population increases strain infrastructures and strained infrastructures lead to stresses on society.

We already are experiencing higher ultraviolet B radiation through our depleted ozone layer and can expect more cancers, more cataracts, and diminishing immunity. Expected changing weather and storm patterns may result in disturbed and diminished agricultural production with malnutrition and famine on a grandiose scale; diseases would migrate and the number of displaced persons would increase greatly.

This book consists of papers presented at a meeting on Global Atmospheric Change and Public Health, held in Washington, D.C., in December 1989. It was sponsored by the Air Resources Information Clearinghouse (ARIC), a project of the Center for Environmental Information, Inc. (CEI), a nonprofit organization in Rochester, New York, and co-sponsored by thirty-two U.S., Canadian and international organizations and agencies.

The conference was the first to bring together in a public forum the health, scientific, policy and information communities to address the issues. The book examines potential public health and health-related impacts on society, communicable diseases, cancer and cataract, immunity, heat effects, respiratory problems and human nutrition.

A special panel presented, for the first time, a discussion on information sources and needs in this cross-disciplinary area where vital material may be hidden in the "gray literature." Also included are papers on research needs and priorities in the public health area.

These proceedings are a cooperative effort of the staff of CEI. Special thanks go to Debra Segura for her assistance with the editing and her role in the production of camera-ready copy. Great credit is due to William Wagner and Carole Beal, the associate editors, for the hours they spent in so capably editing and producing this volume. The conference itself would not have occurred without the organizational skill of Linda Wall, CEI's conference coordinator. The editor appreciates the office services provided by the Center for Environmental Research at Cornell University.

Atmospheric change is here to stay. This book contains a wealth of sound, unbiased information on potential health problems of the next century. We present it as a valuable reference for scientists and those who establish environmental policy.

James C. White
Center for Environmental Research
Cornell University, April 1990

ABOUT THE
CENTER FOR ENVIRONMENTAL INFORMATION

The Center for Environmental Information (CEI) was established in Rochester, New York, in 1974 as an answer to the growing dilemma of where to find timely, accurate and comprehensive information on environmental issues. To meet this need for current and comprehensive information, CEI has developed a multi-faceted program of publications, educational programs and information services. It is a private, nonprofit organization funded by membership dues, fees, contracts, grants and contributions. The Center remains today a Rochester-based organization, but its services now reach far beyond the local community, reflecting the increasing number, scope and complexity of problems affecting the environment.

CEI acts as a catalyst to advance the public agenda toward soundly conceived environmental policies. CEI's communication network provides a link among the scientific community, educators, decision makers and the public, so that informed action follows the free interchange of information and ideas.

CONFERENCE ORGANIZING COMMITTEE

Dr. Richard Ball
Office of Environmental Analysis, U. S. Department of Energy

Mr. Nelson E. Hay
Chief Economist & Director, Policy Analysis,
American Gas Association

Dr. Alexander Leaf
Chairman, Department of Preventative Medicine,
Harvard Medical School

Dr. Janice Longstreth
Executive Director, Division of Science Policy,
Clement Associates

Dr. Gordon J. MacDonald
Vice President and Chief Scientist,
The MITRE Corporation

Mr. Ken Murphy
Executive Director, Environmental & Energy Study Institute

Dr. Ralph M. Perhac
Senior Scientific Advisor, Environment Division,
Electric Power Research Institute

Dr. James C. White
Professor Emeritus, Center for Environmental Research,
Cornell University

CONFERENCE COSPONSORS

Cosponsors

Agency for International Development
American Gas Association
American Lung Association
American Petroleum Institute
American Public Health Association
Centers for Disease Control
Chemical Manufacturers Association
Climate Institute
The Conservation Foundation
Cornell University Center for Environmental Research
Edison Electric Institute
Electric Power Research Institute
Environment Canada
Environmental Defense Fund, Inc.
Environmental and Energy Study Institute
International Institute for Applied Systems Analysis
Motor Vehicle Manufacturers Association
National Aeronautic and Space Administration
National Climate Program Office
National Institute of Allergy and Infectious Diseases
National Institute of Environmental Health Sciences
National Oceanic and Atmospheric Administration
Natural Resources Defense Council
Ontario Ministry of the Environment
Quebec Ministre de l'Environnement
Society of American Foresters
U.S. Council for Energy Awareness
United Nations Environment Programme
United States Department of Agriculture
United States Department of Energy
United States Environmental Protection Agency
World Meteorological Organization
World Resources Institute

Funding Support

American Gas Association
Centers for Disease Control
Electric Power Research Institute
Motor Vehicle Manufacturers Association
National Institute of Environmental Health Sciences
Ontario Ministry of the Environment
U.S. Council for Energy Awareness
United States Environmental Protection Agency
World Resources Institute

OVERVIEW OF GLOBAL ATMOSPHERIC CHANGE

Gordon J. MacDonald

The MITRE Corporation
7525 Colshire Drive
McLean, VA 22102-3481

On November 8, 1989, British Prime Minister Margaret Thatcher addressed the 44th Session of the United Nations General Assembly regarding environmental change. For the first time in the United Nations' history, a prominent national leader discussed in detail the threats to world peace and economic stability flowing from the changes that man is imposing on the global environment. Her thesis is captured in the following quote:

> While the conventional political dangers – the threat of global annihilation, the fact of regional war – appear to be receding, we have all recently become aware of another insidious danger. It is as menacing in its ways as those more accustomed perils with which international diplomacy has concerned itself for centuries. It is the prospect of irretrievable damage to the atmosphere, to the oceans, to the earth itself.

At this conference, we will discuss a most important aspect of this insidious danger, the threat to public health resulting from pervasive changes in the atmosphere. Human activities have brought about massive increases in the concentrations of certain key constituents of the atmosphere: carbon dioxide, methane, nitrous oxide, and the chlorofluorocarbons (CFCs). All of these gases assist water vapor in trapping heat that would otherwise radiate from the earth's surface and atmosphere into space, and thus form a gaseous greenhouse.

The warming effect of the greenhouse, in moderation, is essential to the survival of our species. Yet alterations in the concentrations of the greenhouse gases pose various threats to human health. As the earth warms, periods of extreme heat or cold can be expected to become more frequent and intense, placing severe stress on affected populations. Of great consequence to public health is the intensity and frequency of extreme weather events, such as storm surges, hurricanes, typhoons, etc. Such events not only cause death and destruction but they also produce conditions under which disease can propagate by disrupting food and water supplies.

In addition to warming the planet, the greenhouse gases jointly participate in the destruction of the protective ozone layer in the upper atmosphere. A decrease in stratospheric ozone will expose populations to higher levels of biologically damaging ultraviolet

© 1990 by Elsevier Science Publishing Co., Inc.
Global Atmospheric Change and Public Health
James C. White, Editor

radiation, with consequent adverse impacts on the incidence of skin cancer, eye damage, and the functioning of the immune system. These effects are the principal subjects of this conference. My task is to outline what changes have taken place in the global atmosphere, emphasizing both what we know and what we do not know. Our present knowledge provides a basis for guessing at what the future may bring, but recent experience suggests that predictions must be approached with great caution.

Following the discovery by Molina and Rowland (1974) that CFCs could destroy stratospheric ozone, the Federal Task Force on Inadvertent Modification of the Atmosphere (IMOS, 1975) estimated that the total loss of ozone resulting from projected future uses of CFCs would be 7 percent. By 1979, changes to the input data in computer models had raised the estimated steady state ozone depletion from 7 to 18 percent (NAS, 1979a, 1979b). A further series of changes led in 1984 to an estimate for a total eventual ozone depletion of 5 percent (NAS, 1984). The scientific community, government, policymakers and industry were surprised in 1985 when the British Antarctic Survey published the results of ozone observations taken at Halley Bay (76°S) (Farman, et al., 1985). The survey team had detected a gradual austral springtime decline in stratospheric ozone in the late 1970s that was followed by a precipitous plunge in the early 1980s, as shown in Figure 1. The 40-percent loss of ozone over an area the size of the United States had not been anticipated, largely because computer models had assumed that stratospheric chemistry could be described in terms of homogeneous gas-phase reactions. Heterogeneous reactions on the surface of particles making up polar stratospheric clouds had not been considered. As the effort to predict climate continues, I feel certain that this will not be nature's last surprise.

In the following overview of global atmospheric change, I first describe the compositional changes that are intensifying the greenhouse effect. CFCs are significant greenhouse contributors and are also the source of chlorine which removes stratospheric ozone. Carbon dioxide and methane, the two major greenhouse gases, are important links in stratospheric chemical reactions that lead to ozone destruction. In describing our present understanding of stratospheric chemistry, I emphasize the origin of the Antarctic ozone hole. This discussion raises the question of what implications the Antarctic ozone has for the stability of stratospheric ozone over the rest of the globe. I end with a more speculative examination of future greenhouse warming and the expected intensity and frequency of extreme events.

Changing Atmospheric Composition

Carbon dioxide

The modern era of climate-related atmospheric chemistry began in 1958 when David Keeling commenced his highly precise measurements of carbon dioxide at Mauna Loa, Hawaii, as part of the activities connected with the International Geophysical Year. In 1958, the annual average concentration of CO_2 was 315 ppmv; by 1988, this amount had risen to 351 ppmv. The increase from 315 to 351 ppmv over a 30-year period corresponds to the addition of 76.4 gigaton of carbon to the atmosphere. Currently, the exponential rate of increase of atmospheric carbon dioxide is 0.5 percent per year (see Figure 2). A further feature of the variation of atmospheric CO_2 concentration is its seasonal oscillation, which is due to terrestrial biological activity concentrated mostly in the Northern Hemisphere. Atmospheric CO_2 decreases during the growing season (May-October) and increases through respiration during the colder months. At Mauna Loa, the peak-to-peak amplitude is about 6 ppmv, or 12.7 Gt of carbon. The amplitude of the seasonal oscillation increases at higher latitudes and decreases toward the equator.

The history of the variation of carbon dioxide at high latitudes can be extended backward in time by analyzing the composition of air trapped during the formation of glacial ice in Greenland and Antarctica. Because the conversion of snow to ice takes at least a decade, the dating of past air samples cannot be very precise. Nevertheless, trapped air bubbles give the composition of the atmosphere, averaged over a few years, for the past two centuries. The historical variation of CO_2 derived from air bubbles trapped in glaciers is shown in Figure 3. The preindustrial concentration of atmospheric CO_2 lay in the range of 270 to 280 ppmv and had been at about that level since the end of the last glacial period. If a preindustrial concentration for the year 1740 is taken as 277 ppmv, then about 80.7 Gt of carbon were added to the atmosphere in the 218 years between 1740 and 1958. This amount contrasts with the net addition of 76.4 Gt of carbon during the last 30 years.

Using an ice core from Vostok, Antarctica, a joint French-Soviet team extended the record of atmospheric carbon dioxide farther back into glacial times, reaching past the last interglacial period ($\sim$125-140,000 years ago). Their research shows that, during times of maximum ice coverage in the Northern Hemisphere, the atmospheric CO_2 concentration reached a minimum value of about 180 ppmv (Barnola et al., 1987), as is shown in Figure 4. During the preceding interglacial warm period, the atmospheric CO_2 concentration reached 290 ppmv, a level comparable to preindustrial concentrations.

At the onset of the last two interglacial periods, the atmospheric carbon dioxide concentration increased by about 100 ppmv on a time scale of 10,000 years as a result of, as yet, poorly understood natural processes. Since the industrial revolution began, sufficient carbon dioxide has been added to the atmosphere to increase the concentration by about 70 ppmv on a time scale of only 250 years. During the last 30 years alone, the CO_2 concentration has increased by about 35 ppmv. The recent increases are smaller on a percentage basis (25 percent over 250 years) than during glacial periods (60 percent over 10,000 years). Yet these comparisons are significant; changes at smaller concentrations have a greater proportional impact on the radiative balance than do changes at higher concentrations, because certain CO_2 absorption bands (15 μm) become saturated at higher carbon dioxide concentration levels and contribute proportionally less to the greenhouse effect than at lower concentrations.

The coincidence of the dawning of the industrial-agricultural revolution with the onset of major changes in atmospheric composition signifies that man's activities play an important role in these changes. Three kinds of human activity release carbon dioxide to the atmosphere in a major way: burning fossil fuels, converting tropical and other forested areas to alternative uses, and certain manufacturing processes, particularly the making of cement. The production of one metric ton of cement releases 0.136 metric tons of carbon as CO_2; this figure does not include the fuel burned in calcining the limestone. When forested areas are cleared and converted through use to lands that have smaller inventories of carbon stored in trees, surface litter, and soil, there is a net release of carbon to the atmosphere in amounts that can be large but are often unknown. Also, for every cubic meter of timber burned, about 0.26 metric tons of carbon are released as CO_2. Releases from the third major CO_2 source, fossil fuels, are very large, with current annual emissions exceeding 5.6 billion tons of carbon.

The fossil fuel sources of carbon dioxide can be estimated from records of global industrial activities now collected by the United Nations (Keeling, 1973; Rotty, 1987). The accuracy of estimates of CO_2 emissions from fossil fuels depends on the reliability of information in three areas: quantities of fuel used, the carbon content of each fuel, and the fraction of the fuel that is actually burned. The reliability of the relevant data improved markedly after the oil crisis of 1973, which greatly increased international interest in energy management.

Figure 5 illustrates the historical variation in the amount of carbon added to the atmosphere through the burning of fossil fuels. Fuel combustion over the last 128 years can be separated into four major periods, the first three of which had a distinct rate of exponential growth. Between 1860, when the United Nations data set begins, and 1913, fossil fuel use increased at an average exponential rate of 4.3 percent per year. Coal was the dominant fossil fuel during that period, though large and uncertain amounts of wood were burned. With the advent of World War I, the global growth rate of energy use slowed down markedly to 1.5 percent per year. This slower average rate of growth persisted through the world turbulence of the 1920s, the depression of the 1930s, and World War II. During the 23-year period beginning in 1950, rapid growth in energy use resumed, and oil overtook coal as the dominant fuel. The oil price shock of 1973 began a period of tumult in the world energy business. After 1973, CO_2 emissions leveled off (see Figure 6), but began rising again until 1980, when they dropped in the wake of the Iraq-Iran War. Lower oil prices began to drive emissions up beginning in 1983, with a subsequent growth rate of almost 4 percent per year.

Figure 6 shows that, although the increase in atmospheric carbon dioxide concentration paralleled fossil fuel emissions over much of the 30-year period beginning in 1958, the atmosphere did not respond to the drop in emissions beginning in 1980. The continued monotonic rise in atmospheric CO_2 concentration, despite reduced emissions, indicates that other sources of carbon dioxide, particularly the biosphere, participate in a major way in determining the carbon content of the atmosphere.

Methane

Methane is the most abundant reactive trace gas in the atmosphere; it is also a strong infrared absorber. During its approximate ten-year lifetime, a methane molecule is equivalent in greenhouse warming capacity to 30 carbon dioxide molecules. After being oxidized by the extremely reactive hydroxyl radical into long-lived carbon dioxide, the converted carbon continues to act as a greenhouse absorber. Methane is the source of about one-half the hydrogen and water in the stratosphere, and is thus a major determinant of the stratosphere's chemical and thermal regimes.

In 1988, the global methane concentration was 1.7 ppmv. Various studies have established that atmospheric methane is currently increasing at a rate of 1 percent per year, or about twice the rate at which carbon dioxide is increasing (Rasmussen and Khalil, 1981; Fraser et al., 1984; Blake and Rowland, 1988). A reliable record for methane measured directly from the atmosphere is available only for the years 1978-1989; the quality of earlier measurements is inferior. Rinsland et al. (1985) re-analyzed solar absorption spectra taken on Jungfraujoch in 1951 and deduced a methane concentration for that year of 1.14 ppmv. This measurement is consistent with an exponential increase of methane at the rate of 1.08 percent per year. However, as noted, there is no extended time series for atmospheric methane concentration comparable to that obtained by Keeling for carbon dioxide.

Evidence from the analysis of gases trapped in ice from both Greenland and Antarctica shows that the atmospheric methane concentration doubled during the last 200 years, but remained at a level between 0.6 and 0.8 ppmv over the previous 3,000 years (see Figure 7). The Vostok ice core data (Figure 4) extend the record for methane back 160,000 years to the time of maximum ice coverage. Methane concentrations during glacial peaks were low, about 0.35 ppmv, or almost one-half the value observed during interglacial times (Stauffer et al., 1988). The Vostok ice core data show that methane levels increased from 0.32 ppmv to 0.62 ppmv between the end of the last glaciation and the subsequent interglacial period, paralleling shifts in the CO_2 concentration.

Methane is the most abundant organic gas in the atmosphere; its current concentration of 1.7 ppmv corresponds to 3.61 Gt of carbon. Calculations based on the photochemistry of methane yield an atmospheric residence time for this gas of 8.1 to 11.8 years (Cicerone and Oremland, 1988), which implies a quasi-steady state source of 0.4 to 0.59 Gt of carbon per year. At present, the amount of methane is increasing at a rate of about 0.036 Gt per year.

Isotopic ratios aid in the determination of the sources of methane. Because methane released from "old" deposits will not contain any ^{14}C, which is generated by cosmic rays, nuclear bomb tests, and nuclear reactors, an analysis of the $^{14}CH_4/^{12}CH_4$ ratio provides a measure of the amount of methane that is of relatively recent biological origin. Lowe et al. (1988) estimated that only 68 percent of current methane sources are biogenic. This estimate has recently been revised by Manning et al. (1989) to 74 percent. Wahlen et al. (1989) found that, at the end of 1987, 21 ± 3 percent of atmospheric methane sources were derived from fossil carbon sources. Given the 20- to 30-percent range for fossil carbon in atmospheric methane, modern biogenic sources must annually release between 0.28 and 0.47 Gt of carbon as methane.

Conventional discussions on the origin of the increase in atmospheric methane focus on enteric fermentation in animals, rice paddies, termites, biomass burning, landfills and other sources that tend to be concentrated in the tropics and mid-latitudes. However, given the observed geographic distribution of atmospheric methane, a substantial fraction must have a high-latitude source. Methane concentrations are highest at high latitudes, drop off toward the equator, and remain constant in the Southern Hemisphere. Possible northern sources of CH_4 include releases from natural wetlands and bogs in the boreal forests of Siberia and Canada, releases from methane clathrates in coastal and Arctic settings, and emissions from extensive development of gas fields and coal mines, particularly in Siberia. The increase in biogenic methane releases could result from changes in the abundance of wetlands, or from the thermal stimulation of methane-generating bacteria as a result of global warming.

Chlorofluorocarbons (CFCs)

Unlike methane and carbon dioxide, there are no known natural sources for CFCs such as $CFCl_3$, CF_2Cl_2 and CH_3CCl_3. Industrial production of such halocarbons began recently, in the 1930s, yet they are the most rapidly increasing species in the atmosphere. Essentially inert in the troposphere, CFCs undergo photodecomposition in the stratosphere.

The most abundant chlorine- and fluorine-containing halocarbon is CF_2Cl_2 (CFC-12), which has an atmospheric lifetime of about 111 years and a current concentration of about 0.4 ppbv – roughly 400 times less than the concentration of methane. Despite this small concentration, CFC-12, like other halocarbons, is a potent infrared absorber; one molecule has the equivalent greenhouse impact of 10,000 CO_2 molecules. The rate of increase of atmospheric CFC-12 is about 5.1 percent per year (Prinn, 1988). Worldwide, CFC-12 is primarily used as a refrigerant and as a propellant in aerosol cans.

The next most abundant chlorofluorocarbon is $CFCl_3$ (CFC-11), which is used both in blowing foams and as a propellant in aerosol cans. The atmospheric concentration of CFC-11 has paralleled that of CFC-12 and is currently at a level of about 0.11 ppbv; however, its atmospheric lifetime is less than that of CFC-12, about 74 years (Prinn, 1988). Methyl chloroform (CH_3CCl_3) is primarily used as a solvent for degreasing; its atmospheric concentration is rapidly growing at about 6 percent a year, despite a short lifetime of about six years. Methyl chloride (CH_3Cl) is the most abundant natural halocarbon, with a concentration of 0.61 ppbv, but has a short atmospheric lifetime of only 1.5 years.

Chemistry of Stratospheric Ozone

Stratospheric ozone is formed when ultraviolet radiation from the sun penetrates the atmosphere. Wavelengths shorter than 242 nm split an oxygen molecule into two oxygen atoms, one of which combines with an oxygen molecule to form ozone. The peak ozone concentration (about 1 ppmv) is found 15 to 30 km up in the stratosphere. The number density of ozone drops off not only at lower altitudes because a smaller amount of hard ultraviolet radiation penetrates, but also at higher altitudes because the number of oxygen molecules decreases. Ozone absorbs solar radiation at wavelengths greater than 242 nm, and in this process the stratosphere is warmed. Ozone is removed over temperate and tropical regions by a reaction with an oxygen atom, which gives two oxygen molecules.

The total concentration of organically bound chlorine in the troposphere has increased rapidly since regular measurements began about 15 years ago. The inferred preindustrial concentration of CH_3Cl, the only natural organic compound containing chlorine, was about 0.6 ppbv. The concentration rose to 1.0 ppbv in 1965 and has now reached a value of 3.5 ppbv.

Chlorine produced by the photochemical breakdown of chlorofluorocarbons can accelerate ozone loss by reacting with ozone to form chlorine monoxide. The chlorine monoxide then reacts with an oxygen atom to reform chlorine according to:

$$Cl + O_3 \rightarrow ClO + O_2$$

$$ClO + O \rightarrow Cl + O_2$$

with a net reaction of:

$$O_3 + O \rightarrow 2O_2$$

Catalytic removal of ozone can continue until the process is brought to a halt when the chlorine enters a reservoir molecule such as HCl, which eventually (in a year or so) is carried from the stratosphere to the troposphere to be rained out. Before eventual removal, each chlorine atom destroys approximately 100,000 ozone molecules.

Although the chlorine cycle originally proposed by Molina and Rowland (1974) is descriptive of accelerated ozone removal over much of the atmosphere, it cannot be used to explain the rapid development of the Antarctic ozone hole. Because Antarctica is an isolated continent forming a high, cold dome that is surrounded by the circum-Antarctic Ocean, its stratosphere is quasi-isolated from the rest of the atmosphere in terms of dynamics. During the Southern Hemispheric winter, a cold and isolated air mass forms in the stratosphere, with a vortex-like circulation that spins toward the east. As the sun's rays first strike the Antarctic stratosphere, the long, nearly horizontal path they travel results in attenuation of the ultraviolet radiation capable of splitting oxygen molecules. If a catalytic process is involved in the formation of the ozone hole, it must depend primarily on chain reactions that do not involve atomic oxygen, because atomic oxygen is formed by the photolysis of O_2 by ultraviolet radiation.

The question of the origin of the ozone hole was decisively answered by the scientific expeditions to Antarctica in 1986 and 1987 (NASA, 1988), which showed that ozone destruction by chlorine was aided by the special meteorological conditions existing over Antarctica. The long polar night produces temperatures as low as $-90°C$ at altitudes of 15 to 20 km. At these low temperatures, stratospheric clouds form. Direct sampling has shown that micrometer-sized particles containing nitric acid and water form at temperatures below –78°C. These droplets could be frozen nitric acid trihydrate, which freezes at a higher temperature than does water. At temperatures below about –87°C, larger water-ice crystals were found. The polar stratospheric cloud particles facilitate the destruction of ozone in a number of ways. The particles provide a surface on which heterogeneous chemical reactions can convert hydrochloric acid and chlorine nitrate into molecular chlorine and nitric acid by reactions such as:

$$HCl + ClONO_2 \rightarrow Cl_2 + HONO_2$$

With the first sunlight of approaching spring, atomic chlorine is released through photolysis, with the nitric acid being absorbed into the drop. The ice particles can remove $HONO_2$ from the stratosphere by settling out. This process is essential for the chlorine catalytic cycle involving the chlorine monoxide dimer to proceed:

$$ClO + ClO + M \rightarrow Cl_2O_2 + M$$

$$Cl_2O_2 + h\nu \rightarrow ClO_2 + Cl$$

$$ClO_2 + M \rightarrow Cl + O_2 + M$$

$$2 \times (Cl + O_3 \rightarrow ClO + O_2)$$

If $HONO_2$ is not removed, $ClONO_2$ acts as a reservoir by tying up chlorine.

Extremely high concentrations of chlorine monoxide have been measured in the Antarctic polar vortex, both from ground and aircraft observations. The striking correlation between the high concentration of chlorine monoxide and the low ozone concentration provides strong support for the ozone-depleting role of the chlorine monoxide's dimer catalytic cycle. The high concentration of chlorine demonstrates the role of man-made CFCs in the growth of the ozone hole.

The formation of polar stratospheric clouds, which are essential to the heterogeneous destruction of ozone, depends critically on the presence of water in the stratosphere. Cloud particles will only form at H_2O concentrations exceeding the saturation vapor pressure. The vapor pressure over ice at temperatures of –80 to –90°C is doubled by an increase in temperature of 4°C. Increases in water concentration and decreases in stratospheric temperature will favor the formation of stratospheric clouds and the destruction of ozone through heterogeneous clouds. Increasing carbon dioxide in the stratosphere will lead to a *decrease* in stratospheric temperature, while an increase in methane will raise the water content of the stratosphere.

Rowland (1988) illustrates the impact of increasing tropospheric methane concentration on stratospheric water content by a simple calculation. Before the start of the industrial revolution, air parcels entering the stratosphere, which were often associated with tropical thunderstorm activity, carried 3 ppmv H_2O and 0.7 ppmv CH_4. Under equilibrium conditions, with each methane molecule providing hydrogen for two water molecules, the steady state concentration of water in the troposphere would have been

4.4 ppmv. Today, with the entering air parcel carrying 3 ppmv H_2O and 1.7 ppmv CH_4, the steady state concentration of water in the stratosphere should be 6.4 ppmv, an increase of 45 percent over the past 200 years. These figures imply that past temperatures would have to have been at least 2°C colder than at present for polar stratospheric clouds to form.

Carbon dioxide mixes throughout the atmosphere and, unlike water vapor, maintains its concentration in the stratosphere. With few infrared absorbers above it, stratospheric CO_2 radiates into space and thus has a cooling effect. Radiosonde observations show lower temperatures both in the low (9-16 km) and high (16-20 km) stratosphere. The cooling of the stratosphere by a few tenths of a degree per decade has two effects on ozone removal: reaction rates involving ozone destruction by atomic oxygen decrease, allowing CO_2 to preserve ozone where the Molina-Rowland reaction is dominant; ozone removal by heterogeneous reactions involving the chlorine monoxide dimer chain is enhanced in polar regions, where cooling favors the formation of cloud particles at lower water vapor pressures.

What Implications Does the Antarctic Ozone Hole Have for Global Ozone?

The very special conditions existing over Antarctica raise the question as to whether a similar phenomenon could take place elsewhere, particularly in the North Polar regions, which are situated much closer to heavily populated regions. In the Arctic polar regions, temperatures currently do not often reach –80 to –90°C. As a result, the larger ice particles that can remove $HONO_2$ do not form, or form much less frequently than in Antarctica. As noted earlier, the removal of $HONO_2$ is essential for the catalytic destruction of ozone through heterogeneous reactions on the surfaces of nitric acid-water particles. The dynamics of the polar vortex in the Arctic differs greatly from that of the Antarctic. The more complicated geography of the north produces atmospheric wave activity that leads to more frequent breakup of the smaller and more irregular Arctic vortex. Despite the fundamental differences between the two polar regions, early evidence, including high concentrations of ClO, indicates that heterogeneous removal of ozone occurs in the Arctic (Solomon et al., 1988; Mount et al., 1988). Observations on wintertime Arctic chemistry are consistent with the ozone trend observed in the Northern Hemisphere, which exceeds the trend that would be anticipated if only the Molina-Rowland homogeneous gas phase reaction were involved (Rowland et al., 1988; NASA, 1988).

The Antarctic ozone hole can lead to lower stratospheric ozone levels throughout the Southern Hemisphere as ozone-depleted air mixes with lower-latitude air. The degree of depletion is limited by the restricted volume of the polar holes and by their short duration (early spring). Because CFCs deposited in the troposphere take roughly ten years to reach the polar stratosphere, the intensity and size of the ozone depletion will grow as chlorine concentration continues to build, particularly in the Arctic. The effects of heightened chlorine concentration will be intensified by the increase of water vapor in the stratosphere due to increased methane in the troposphere, and by further cooling due to higher CO_2 concentrations. Owing to the difficulty of modeling complex heterogeneous reactions that include the combined effects of increased chlorine, methane and water vapor, quantitative predictions of future ozone depletion are not available.

The stratosphere is considerably warmer outside polar regions than over the vulnerable poles. Lower stratospheric temperatures over mid-latitudes are in the -75 to $-70°C$ region, too warm for either nitric acid-water droplets or ice droplets to form. In the near future, ozone depletion over mid-latitudes is unlikely. However, the combined effects of a $10°C$ cooling with a 20-percent increase in water vapor would stimulate cloud particle formation over wider geographical areas. Such conditions can be anticipated in the middle of the next century due to expected increases in carbon dioxide and methane concentrations. The implications of such developments for public health are clearly serious. Mid-latitude ozone is also vulnerable to large volcanic explosions, which could introduce into the stratosphere particles whose surfaces could support heterogeneous processes.

Extreme Events

In a statistical sense, climate is expected weather. For a particular region, certain events will be common or highly probable. These events will lie close to the central tendency, or mean, of the distribution of all weather events. Other types of events will be more extreme and less frequent. The more extreme the event, the lower the probability of its occurrence. The overall distribution of climate parameters defines climate variability for a particular location. Temperature is one such parameter. Most discussions of climate change focus on changes in average temperature, which is a relatively uninteresting parameter for public health considerations. Of greater interest are the frequency and intensity of extreme events such as hot spells and storms. In particular, tropical cyclones rank with earthquakes as major natural causes of loss of life and property. Storm surges and hurricanes can cause difficult public health problems by disrupting food and water supplies as well as sewage systems.

The warming of the atmosphere due to the greenhouse will raise the water vapor content of the troposphere and the energy contained within the atmosphere, principally as a result of increased sea-surface temperatures. Observations of storms indicate that higher sea-surface temperatures are accompanied by intensified winds. A simple thermodynamic model for the intensity of the pressure drop at the center of a storm and the maximum associated winds has been developed by Emanuel (1987). Coupling these results to estimates of sea-surface temperature increases associated with a doubling of atmospheric CO_2 yields a probable increase in the destructive potential of hurricanes of 40 to 50 percent.

The study of extreme events associated with warming climate is just beginning. Early results suggest that both frequency and intensity of extreme events will increase in a greenhouse-warmed world. If these findings hold, they will have major implications for public health.

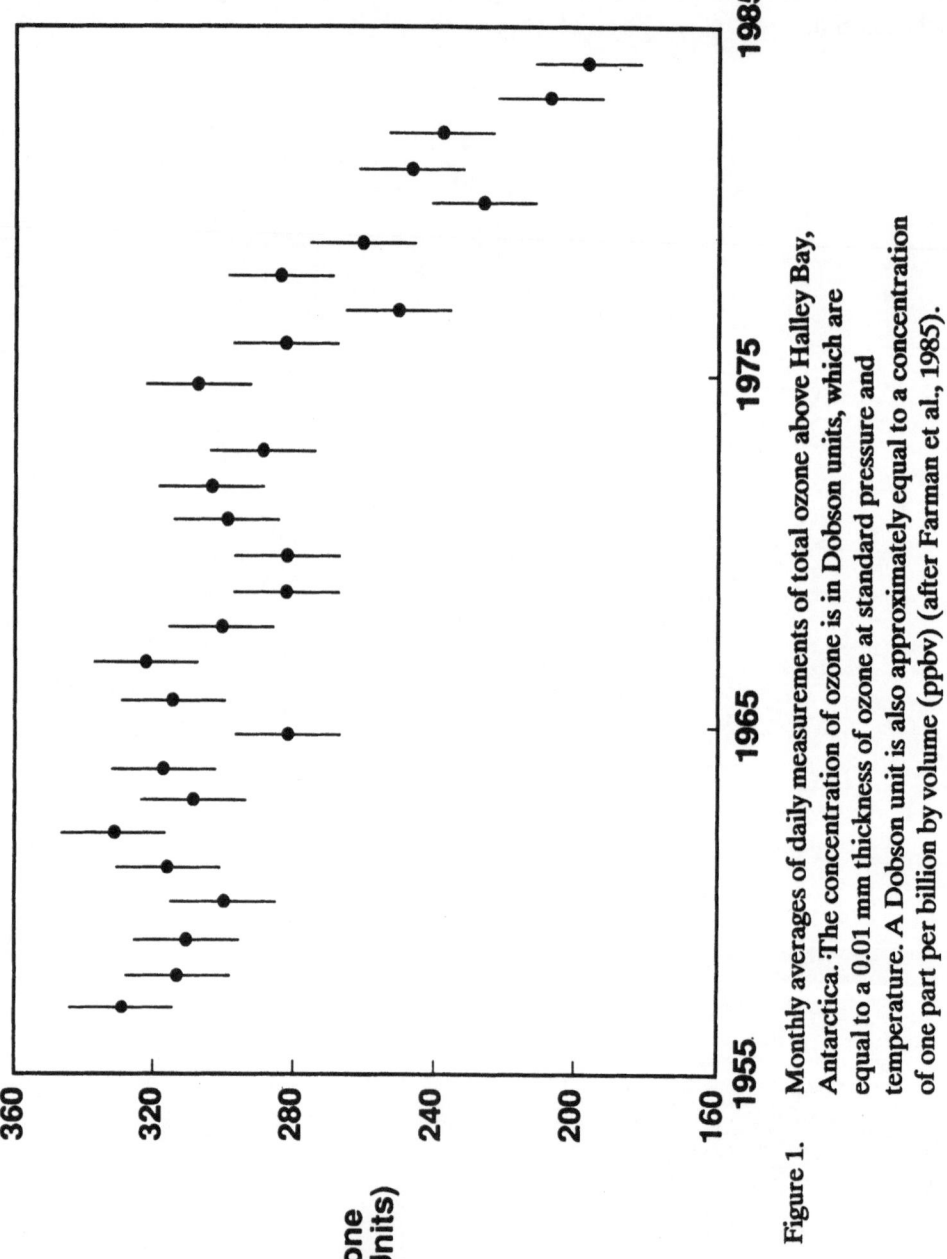

Figure 1.   Monthly averages of daily measurements of total ozone above Halley Bay,
Antarctica. The concentration of ozone is in Dobson units, which are
equal to a 0.01 mm thickness of ozone at standard pressure and
temperature. A Dobson unit is also approximately equal to a concentration
of one part per billion by volume (ppbv) (after Farman et al., 1985).

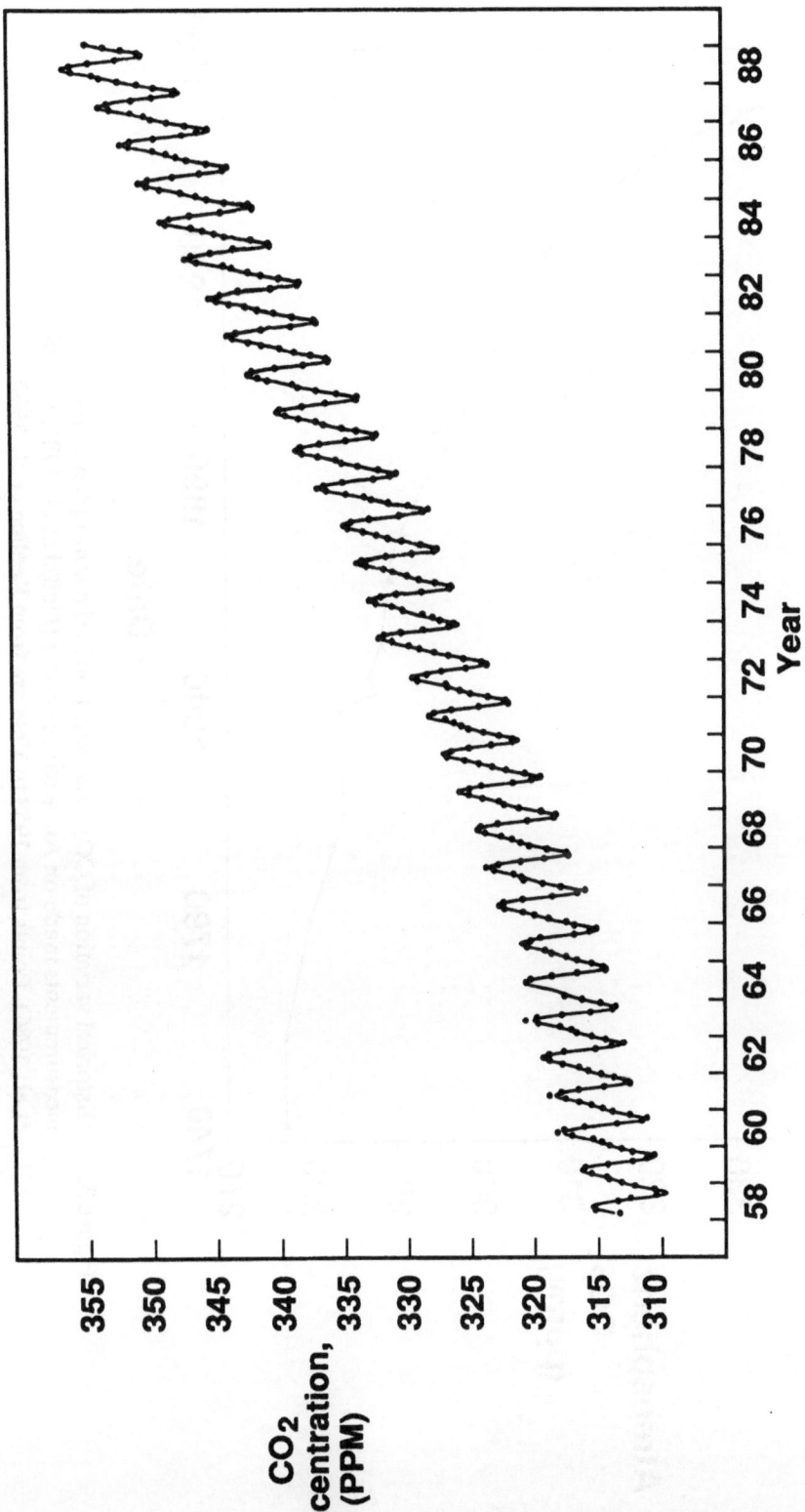

Figure 2. Variation of atmospheric concentrations of carbon dioxide at Mauna Loa
Observatory, Hawaii (after Keeling et al., 1990).

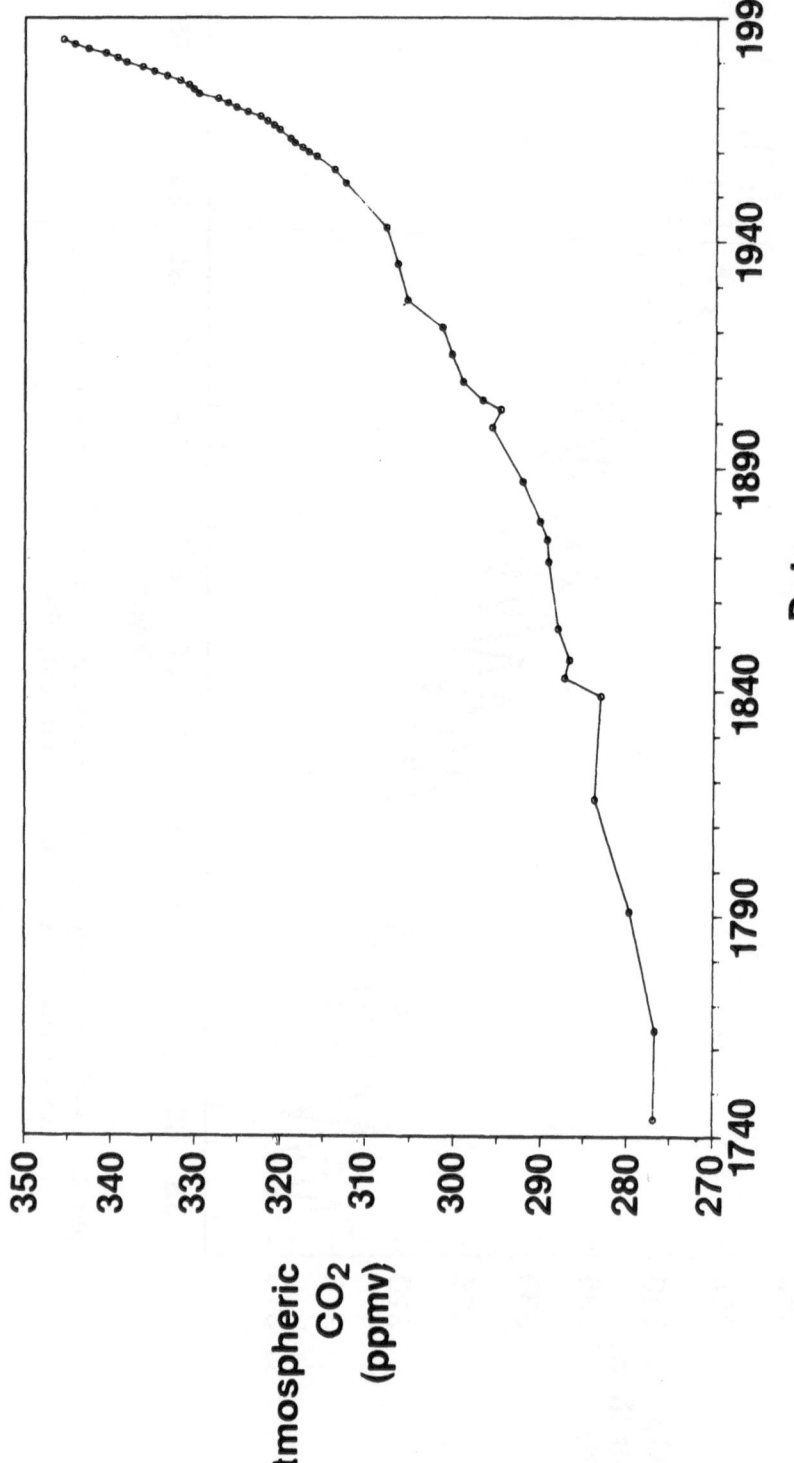

Figure 3.   Historical variation of CO_2 concentration in the atmosphere, from
measurements made on Antarctic ice cores (Neftel et al., 1985; Friedli
et al., 1986). Points from 1958 to 1988 are from Keeling et al., 1990.

Comparison of Glacial Age Variation in Atmospheric CO₂ & CH₄

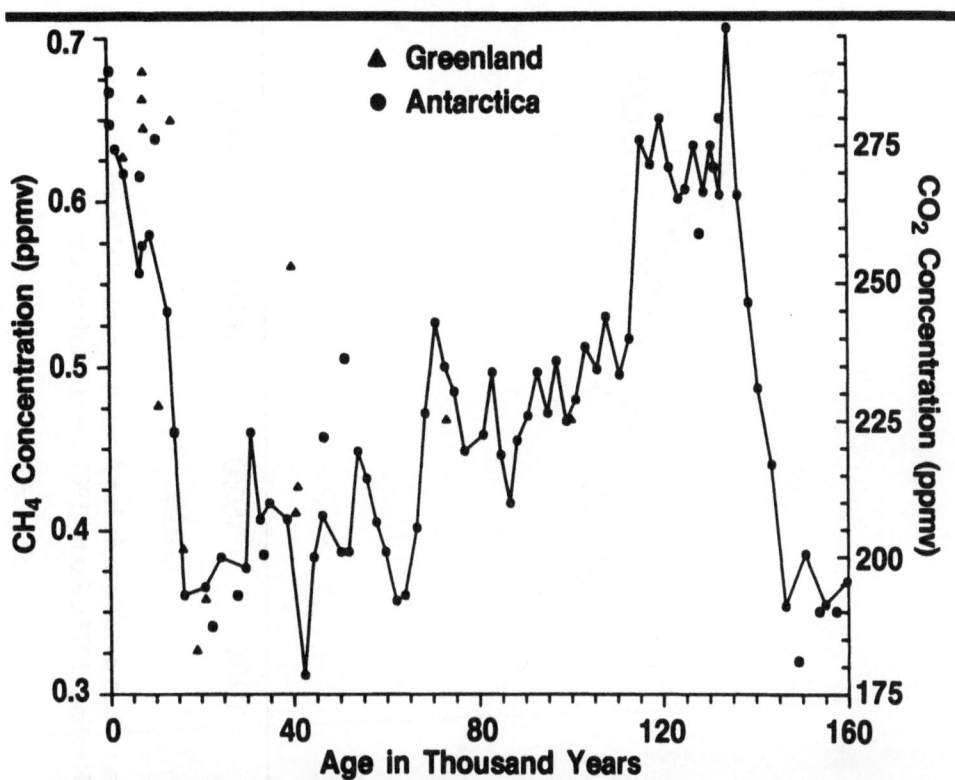

Figure 4.   Variation of methane and carbon dioxide during the past 160,000 years.
The methane data are computed from Stauffer et al. (1985), Stauffer
et al. (1988), Raynaud et al. (1988), and Craig and Chou (1982). The
solid line representing the CO₂ variation in air trapped in the Vostok
ice core is from Barnola et al. (1987).

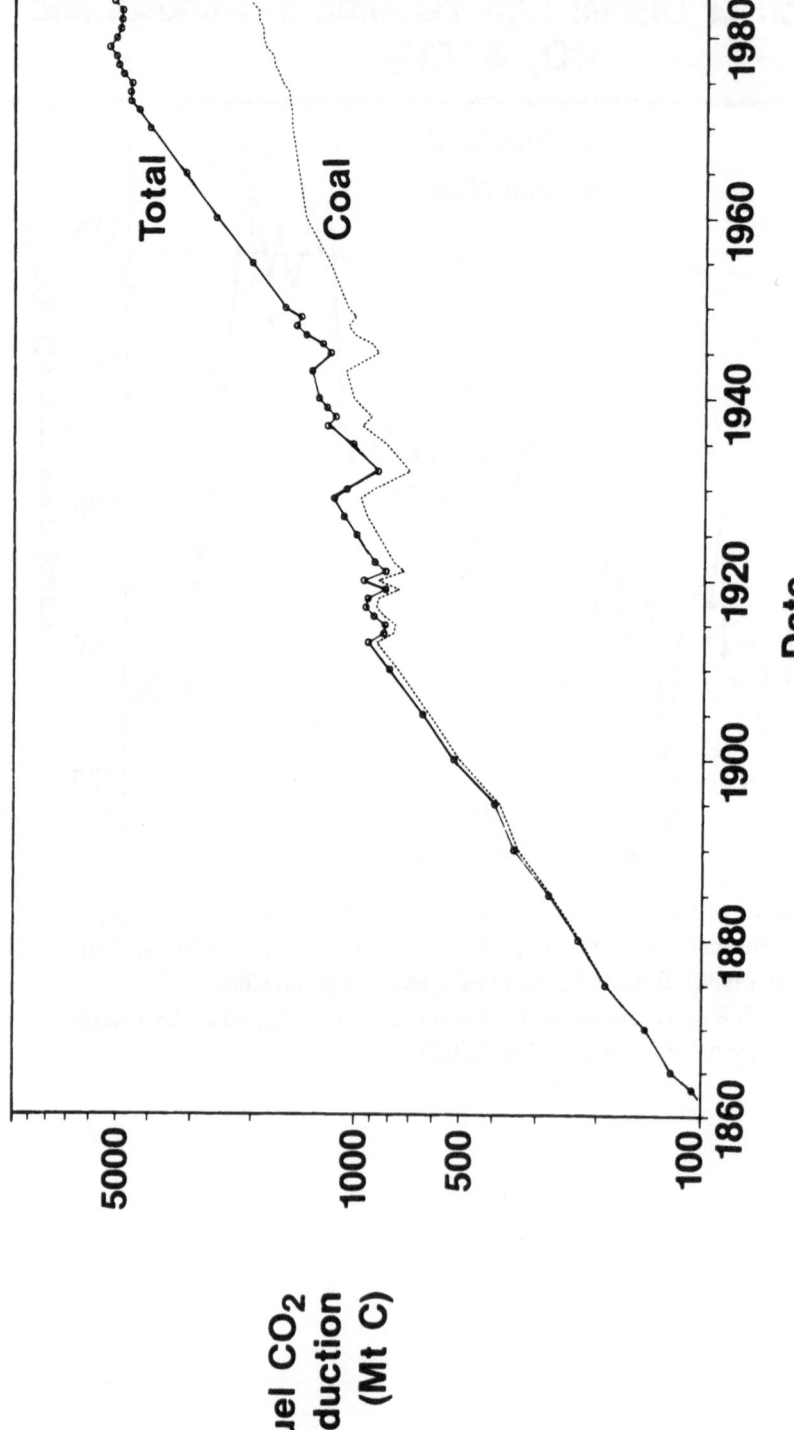

Figure 5.   Historical variation in carbon dioxide emission from the burning of fossil
             fuels. Data are from Keeling (1973), Rotty (1987), and Marland and
             Rotty (1984).

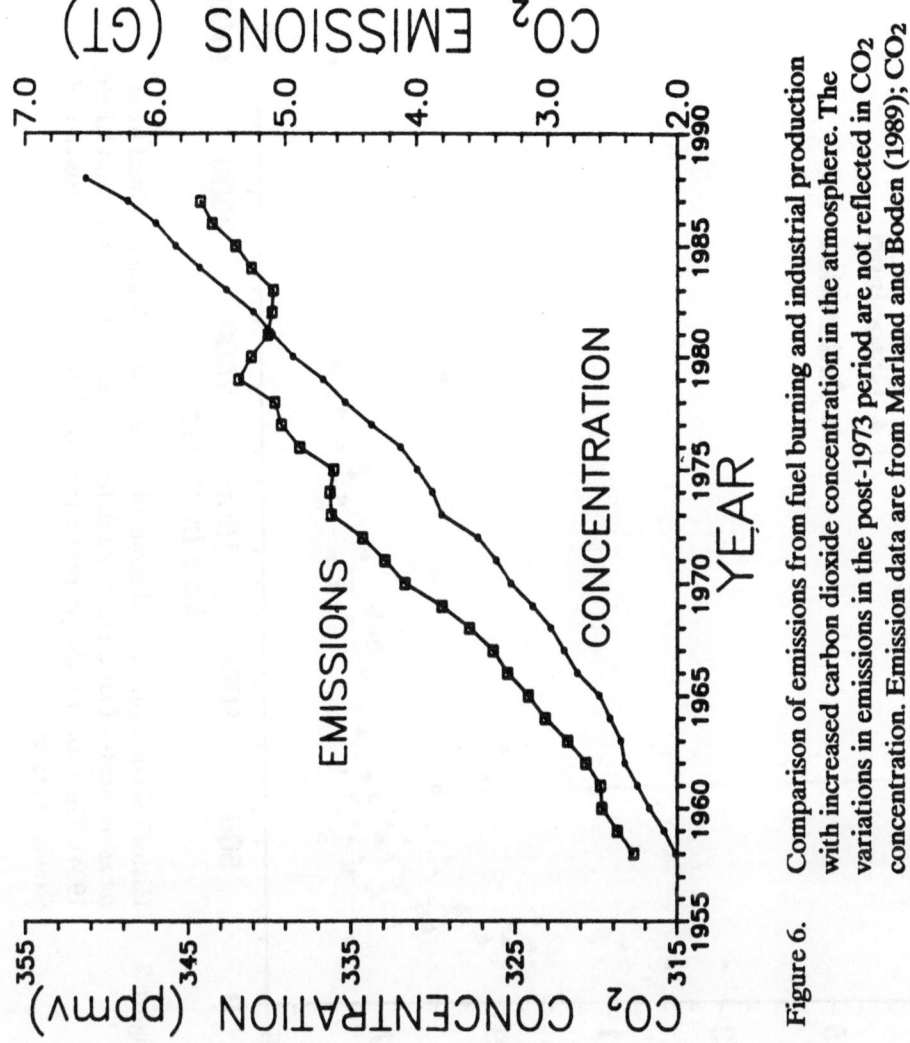

Figure 6. Comparison of emissions from fuel burning and industrial production with increased carbon dioxide concentration in the atmosphere. The variations in emissions in the post-1973 period are not reflected in CO_2 concentration. Emission data are from Marland and Boden (1989); CO_2 data are yearly averages of values shown in Figure 1.

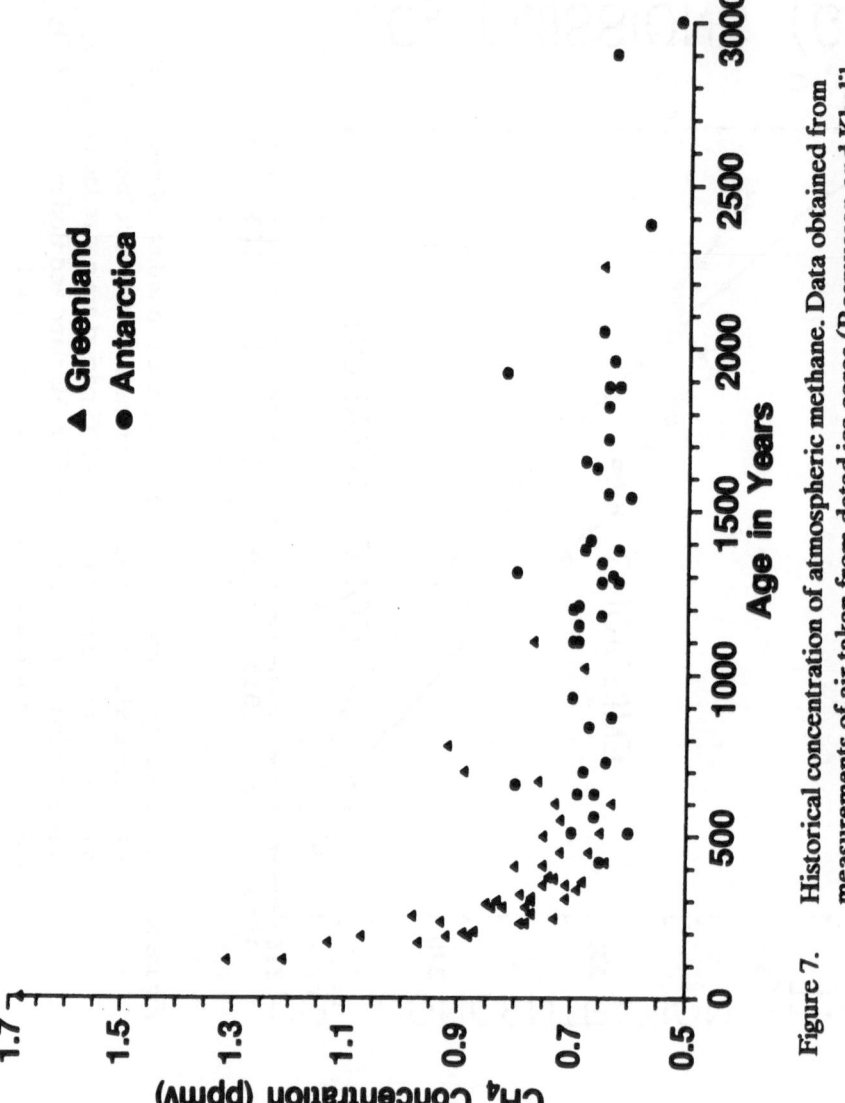

Figure 7.   Historical concentration of atmospheric methane. Data obtained from measurements of air taken from dated ice cores (Rasmussen and Khalil, 1984). The value for the present concentration is taken from Blake and Rowland (1988).

References

Barnola, J., D. Raynaud, Y. Korotkevich and C. Lorius, 1987, "Vostok ice core provides 160,000-year record of atmospheric CO_2," *Nature*, **329**, 408–414.

Blake, D. and F. Rowland, 1988, "Continuing worldwide increases in tropospheric methane, 1978 to 1987," *Science*, **239**, 1129–1131.

Cicerone, R.J. and R.S. Oremland, 1988, "Biogeochemical aspects of atmospheric methane," *Global Biogeochemical Cycles*, **2**, 299–327.

Craig, H. and C. Chou, 1982, "Methane: The record in polar ice cores," *Geophysical Research Letters*, **9**, 1,221–1,224.

Emanuel, K., 1987, "The dependence of hurricane intensity on climate," *Nature*, **326**, 483–485.

Farman, J., B. Gardiner and J. Shanklin, 1985, "Large losses of total ozone in Antarctica reveal seasonal ClO_x/NO_x interaction," *Nature*, **315**, 207–210.

Fraser, P., M. Khalil, R. Rasmussen and L. Steele, 1984, "Tropospheric methane in the mid-latitudes of the southern hemisphere," *Journal of Atmospheric Chemistry*, **1**, 125–135.

Friedli, H., H. Lötscher, H. Oeschger, U. Siegenthaler and B. Stauffer, 1986, "Ice core record of the $^{14}C/^{12}C$ ratio of atmospheric CO_2 in the past two centuries," *Nature*, **324**, 237–238.

IMOS (Federal Task Force on Inadvertent Modification of the Stratosphere), 1975, *Fluorocarbons and the Environment*, Council on Environmental Quality, Federal Council for Science and Technology.

Keeling, C., 1973, "Industrial production of carbon dioxide from fossil fuels and limestone," *Tellus*, **25**, 174–198.

Keeling, C., R. Bacastow, A. Carter, S. Piper, T. Whorf, M. Heimann, W. Mook and H. Roeloffzen, 1990, "A three-dimensional model of atmospheric CO_2 transport based on observed winds: I. Observational data and preliminary analysis," *Journal of Geophysical Research* (in press).

Lowe, D., C. Brenninkmeijer, M. Manning, R. Sparks and G. Wallace, 1988, "Radiocarbon determination of atmospheric methane at Baring Head, New Zealand," *Nature*, **332**, 522–525.

Manning, M., D. Lowe, W. Melhulsh, R. Sparks, G. Wallace and C. Brenninkmeijer, 1989, "The use of radiocarbon measurements in atmospheric studies," *Radiocarbons*, **31** (in press).

Marland, G. and R. Rotty, 1984, "Carbon dioxide emissions from fossil fuels: A procedure for estimation and results for 1950–1982," *Tellus*, **36B**, 232–261.

Marland, G. and T. Boden, 1989, "Carbon dioxide releases from fossil-fuel burning," Statement before Senate Committee on Energy and Natural Resources, July 26, 1989, Washington, DC.

Molina, M. and F. Rowland, 1974, "Stratospheric sink for chlorofluoromethanes: Chlorine atom-catalyzed destruction of ozone," *Nature*, **249**, 810–812.

Mount, G., S. Solomon, R. Sanders, R. Jakoubek and A. Schmeltekopf, 1988, "Observations of stratospheric NO_2 and O_3 at Thule, Greenland," *Science*, **242**, 555–558.

NAS (National Academy of Sciences Panel on Stratospheric Chemistry and Transport), 1979a, *Stratospheric Ozone Depletion by Hydrocarbons, Chemistry and Transport*.

NAS (National Academy of Sciences Committee on Impacts of Stratospheric Change), 1979b, *Protection Against Depletion of Stratospheric Ozone by Chlorofluorocarbons*.

NAS (The National Academy of Sciences Committee on Causes and Effects of Changes in Stratospheric Ozone), 1984, *Causes and Effects of Changes in Stratospheric Ozone: Update 1983*.

NASA, 1988, *Present State of Knowledge of the Upper Atmosphere 1988: An Assessment Report*, NASA publ. 1208.

Neftel, A., E. Moor, H. Oeschger and B. Stauffer, 1985, "Evidence from polar ice cores for the increase in atmospheric CO_2 in the past two centuries," *Nature*, **315**, 45–47.

Prinn, R., 1988, "How have the atmospheric concentrations of the halocarbons changed?" in *The Changing Atmosphere*, ed. F. Rowland and I. Isaksen, New York: J. Wiley, 33–48.

Rasmussen, R. and M. Khalil, 1981, "Atmospheric methane (CH_4): Trends and seasonal cycles," *Journal of Geophysical Research*, **86**, 9826–9832.

Rasmussen, R. and M. Khalil, 1984, "Atmospheric methane in the recent and ancient atmospheres: Concentrations, trends, and interhemispheric gradient," *Journal of Geophysical Research*, **89**, 11,599–11,605.

Raynaud, D., J. Chappellaz, J. Barnola, Y. Korotkevich, and C. Lorius, 1988, "Climatic and CH_4 cycle implications of glacial-interglacial CH_4 change in the Vostok ice core," *Nature*, **333**, 655–657.

Rinsland, C., J. Levine and T. Miles, 1985, "Concentration of methane in the troposphere deduced from 1951 infrared solar spectra," *Nature*, **318**, 245–249.

Rotty, R., 1987, "A look at 1983 CO_2 emissions from fossil fuels (with preliminary data for 1984), *Tellus*, **39B**, 203–208.

Rowland, F., 1988, "Some aspects of chemistry in the springtime Antarctic stratosphere," in *The Changing Atmosphere*, ed. F. Rowland and I. Isaksen, New York: J. Wiley, 121–140.

Rowland, F., N. Harris, R. Bojkov and P. Bloomfield, 1988, "Statistical error analysis of ozone trends — Winter depletion in the Northern Hemisphere," *Proceedings of the Quadrennial Ozone Symposium — 1988*, ed. R. Bojkov and L. Fabian, Deepak Publishing.

Solomon, S., G. Mount, R. Sanders, R. Jakoubek and A. Schmeltekopf, 1988, "Observations of the nighttime abundance of OClO in the winter stratosphere above Thule, Greenland," *Science*, **242**, 550–555.

Stauffer, B., G. Fischer, A. Neftel and H. Oeschger, 1985, "Increase of atmospheric methane recorded in Antarctic ice core," *Science*, **229**, 1,386–1,388.

Stauffer, B., E. Lochbronner, H. Oeschger and J. Schwander, 1988, "Methane concentration in the glacial atmosphere was only half that of the preindustrial Holocene," *Nature*, **332**, 812–814.

Wahlen, M., N. Tanaka, R. Henry, B. Deck, J. Zeglen, J. Vogel, J. Southon, A. Shemesh, R. Fairbanks and W. Broecker, 1989, "Carbon-14 in methane sources and in atmospheric methane: The contribution from fossil carbon," *Science*, **245**, 286–290.

IMMUNE SYSTEM AND ULTRAVIOLET LIGHT

Raymond A. Daynes

Department of Pathology
University of Utah School of Medicine
Salt Lake City, Utah 84132

Introduction

The primary function of the mammalian immune system is to allow the vertebrate organisms to coexist with an extremely diverse range of microbial agents having the potential to cause disease. To function, the mammalian immune system requires the input of many distinct cell types. It is a democratic system with many checks and balances in that no single cell type is allowed to carry out effector functions without controlling influences being exerted by a variety of other cell types. The various cell types which collectively comprise the mammalian immune system have to interact and be in constant communication with one another. The tremendous diversity in potentially infectious agents or spontaneously transformed cells that can infect or arise during the lifespan of an individual has provided the selective pressures necessary for the immune system to evolve a variety of different effector mechanisms. It is clear that no single type of immune effector mechanism can handle the diversity of infectious agents which can cause disease.

The central player in the adaptive mammalian immune system is a cell type which is called the T lymphocyte. There is a tremendous diversity in the types and specificities of T cells, predicated on the need of this cell type to recognize and respond against foreign structures while avoiding reactivity against self components.

T cells bear antigen-specific receptors on their cell surfaces. These receptors have evolved the capacity to recognize and be activated by foreign peptide antigens when presented in the context of a self major histocompatibility molecule (MHC). This means that the T-cell receptor is incapable of recognizing nominal antigens, and can differentiate between a foreign peptide and a self-peptide when either are presented in the context of an appropriate MHC-restricting element.

The T-cell receptor arises during the differentiation events taking place within the thymus during T-cell ontogeny. New molecular approaches have provided the experimental means to elucidate how the alpha/beta or the gamma/delta chains of this receptor, a disulfide-linked heterodimer, actually arise from a series of recombination events within individual precursor T cells while resident within the thymus. This is complemented by both negative and positive selection events that are essential for establishing the appropriate restriction, plus a parallel loss of self-reactive potential. This complex process of elimination leaves individual T cells exported from the thymus with the capacity to recognize foreign structures that were never previously encountered. During this same differentiation process, T-cell-bearing receptors having the capacity to recognize anything that is deemed "self" are physically eliminated. This is an important aspect to maintaining a homeostatic balance with other tissues since most types of immunologic effector responses are quite tissue destructive. It would be counterproductive for a mammalian host to have functional T cells that possessed the capacity to recognize and respond to self molecules.

The T-cell receptor has evolved to recognize peptides that can exist on all cell types when the peptides are presented with a self-MHC molecule. MHC encoded molecules are of two basic types. These include molecules which are constitutively expressed on all of our cell types and are called MHC Class I molecules. The genes that encode these structures are highly polymorphic, providing tremendous diversity to individuals within a given species. There also exist MHC Class II molecules. These are also cell-associated polymorphic molecules which exhibit greater selectivity in tissue distribution than the Class I molecules and are found primarily on cell types associated directly with the immune system.

These two types of MHC encoded molecules are important because they possess the capacity to present to T cells different types of antigenic peptides. Antigenic peptides or foreign molecules that are encountered exogenously are endocytosed or phagocytosed by many cell types. Exogenously-presented antigenic substances remain in vesicles, and are partially degraded to peptide fragments during the recycling process. These fragments are then re-expressed on the plasma membrane of the phagocytic cell in association with a MHC Class II molecule. Agents, like bacterial products or other microbial agents, can be phagocytosed, processed and presented to the immune system by these antigen-presenting cells, in the context of MHC Class II molecules.

Protein molecules which arise endogenously, including the molecules associated with possible antigenic disparities that exist in tumors, or foreign molecules present on intracellular parasites, are presented to the immune system via a mechanism which is distinct from that described above. It appears that, during their synthesis and normal intracellular catabolism, peptide fragments of these molecules become associated with MHC Class I molecules during the assembly of the heavy chain of MHC Class I molecules with β_2 microglobulin.

There are many different types of T cells. Some of these distinct T-cell types can be differentiated by their expression of cell-surface molecules. Specifically, the T cells from most species of mammals investigated are CD4$^+$ or CD8$^+$. The CD4$^+$ subpopulation of T cells represents the cells which exhibit a preference for recognizing antigen peptides when they are presented in the context of MHC Class II molecules. In contrast, the CD8$^+$ T cells recognize antigen peptides that are presented in the context of an MHC Class I molecule.

The presence of a very sophisticated mechanism associated with T-cell recognition is an extremely important aspect to a functioning immune system. The fundamental ability to recognize self from non-self involves, at the level of the T cell, both the existence of an antigen-specific T-cell receptor and also the requirement that the antigen be presented appropriately in the context of one or more of the host's MHC molecules.

Subsequent to antigen recognition, the T cells become activated. This leads to both an expansion in numbers of the antigen-specific T cells (clonal expansion) and the production and secretion, by the activated T cell, of a large number of biologically active molecules that are collectively referred to as lymphokines.

The repertoire or pattern of lymphokine species that are made by an individual T cell following its activation by antigen would be extremely important to the development of immunity, and any changes in that pattern could influence the type of immunologic response that ultimately dominates following an encounter with particular antigen. As it would be detrimental to a host to lack the capacity to mount an immunologic response, a failure to mount the appropriate type of immunologic response following infection with an exogenous agent could be just as detrimental.

Experimental Photocarcinogenesis

Our laboratory became interested in the effects of ultraviolet radiation on the immune system during studies of tumors induced by this physical agent. It has been known for many decades that ultraviolet radiation represented a very potent carcinogen, a carcinogen that we know is capable of causing cancer in humans. Studies by photobiologists have provided a tremendous amount of information through their animal studies and have elucidated such aspects as energy requirements, the appropriate wavelengths of light which are capable of causing tumors in rodents, and how this experimental information might relate to tumors induced by ultraviolet radiation in humans.

The daily exposure of shaved mice to the effects of ultraviolet radiation, particularly wavelengths within the UVB range, ultimately results in the development of neoplasms on the exposed surfaces. As a group, these cancers possess a number of interesting characteristics. Studies into transplantation biology of experimental UVR carcinogenesis have determined that it takes a defined amount of UVB energy in a given strain of animals to induce cancer, and that the majority of these progressively growing tumors, when transplanted into normal genetically identical recipients, are unable to grow within the transplant recipients. Normal animals that are exposed to a fraction of the overall dose of UVR needed to generate neoplasms are fully susceptible to tumor implantation and allow these tumors to grow progressively to the recipients' ultimate death. Further, it is possible to demonstrate that, once an animal has received a sufficient dose of UVR to generate tumor susceptibility, it rarely, if ever, recovers from the susceptibility-inducing effect.

We now appreciate that the ability of an animal to either reject or accept UVR-induced tumors is mediated by the immune system. This leads to the logical conclusion, therefore, that the immune system must be capable of being modulated by the effects of UVR. Therefore, in addition to UVR functioning as an extremely potent carcinogen, it also appears that this physical agent is able to modify or depress the capacity of exposed animals to mount the types of immune responses that are necessary to effectively deal with immunogenic tumors. The data generated from these types of experiments are quite striking in that almost every animal that is exposed to UVR demonstrates progressive growth to implanted UVR-induced tumors, whereas normal syngeneic animals fully reject challenge with the same tumor.

Local UVR Effects on the Immune System

What does ultraviolet radiation do to cause the changes in an animal's cellular immune system whereby animals gain a marked susceptibility to UVR-induced tumors? Further, is this change facilitated by some depression in their immune function? If we take a careful look at the wavelengths of solar energy that cause tumors as well as modify immune function in animals, both are within the UVB area of the spectrum (280-320 nm). By assessing the depth of UVB penetration into the skin, a conclusion can be drawn that any systemic changes in immune function caused by this physical agent must be mediated by effects that are probably secondary to events taking place within the skin directly. More specifically, the major direct influences of UVR are confined to the uppermost layer of the skin (the epidermis), plus a small amount of underlying dermal tissue.

Within the skin of mammals there are many resident cell types which originated in other tissues. The macrophages found residing and recirculating through the dermis, and the Langerhans cells which reside within the epidermis, both represent cell types of bone marrow origin that are continually being repopulated from new precursors which arise in an individual's bone marrow. Both dermal macrophages and Langerhans cells are important to a functional immune system by providing a means for antigen presentation. These cells may also be involved in some of the immune effector responses that take place in skin. The exposure of normal skin to the effects of UVR causes rapid changes in phenotypic properties of Langerhans cells within the epidermis. By employing a stain that marks these highly dendritic cells (an antibody with specificity for MHC Class II molecules), it is possible to easily observe the abundance of this cell type in the normal epidermis. Exposure of skin to low-dose UVR *in vivo* results in a loss of recognizable Langerhans cells at the site of UVR exposure. These cells lose both their ability to express MHC Class II molecules, and their ability to effectively present antigen to T cells. Loss of antigen presentation function has been observed with a variety of chemical antigens. A similar type of functional loss in antigen-presenting cell activity takes place when infectious diseases, like viruses, are used to challenge UVR-exposed recipients.

There also exist other cell types in the skin which are directly affected by UVR. These include the endothelial cells which comprise the intricate microvasculature found within the skin of mammals. The microvasculature structures, within skin that has been chronically exposed to ultraviolet radiation, exhibit tremendous long-lasting changes from normal. Normal integrity of the microvasculature within the skin is not only important to maintaining our capacity to effectively function as homeotherms, but is also very important to the maintenance of the integrity of an immune system that relies quite heavily on the ability of the various cell types that are involved in the system to move freely into and out of this tissue.

Another bone marrow-derived cell found residing in the skin is the mast cell. These cells are filled with granules that contain vasoactive and other bioactive substances. Mast cells represent the cell type which is responsible, following degranulation, for the visible reddening and swelling which accompanies many skin allergies. UVR facilitates the degranulation of the mast cell.

One of the most immunologically important cell types resident within skin is the keratinocyte. A few years ago, it was established that the epidermal keratinocytes, following their stimulation with a wide variety of different substances, are capable of producing molecules that are identical to the substances that were previously believed to be manufactured exclusively by cells (macrophages and lymphocytes) of the immune system. The initial observations in this system were made with a cytokine which is now appreciated to be interleukin-1 (IL-1). IL-1 has a long history and historically has gone by a number of different names. Originally described as being an endogenous pyrogen back in the 1950s, it now turns out that IL-1 is probably a major contributor toward the thermal regulation in warm blooded animals. IL-1 also represents the main protein molecule capable of causing fever.

Interleukin-1 is capable, depending upon the receptor-bearing target organ with which it interacts, of causing a wide variety of distinct biologic changes in a host. This substance is capable of stimulating liver cells to change their synthesis of proteins, resulting in the development of a classical acute-phase response. IL-1 can also cause cachexia by its ability to increase the rate of muscle protein catabolism. Further, IL-1 exposure is capable of increasing collagenase production by macrophages and has also been implicated in the pathophysiology of the disease rheumatoid arthritis. It has recently been determined that IL-1, by its capacity to directly stimulate both the hypothalamus and the pituitary gland, can facilitate an increase in the secretion of ACTH. This peptide functions to stimulate the adrenals to elevate their production of anti-inflammatory glucocorticoids. Analysis of plasma from animals that have been exposed to UVR establishes that the total array of acute-phase reactants is elevated within a few hours after a single exposure. This finding documents that UVR, probably through an ability to promote a major increase in IL-1 synthesis by both macrophages and epidermal keratinocytes, is capable of promoting systemic alterations within a host.

Systemic Alterations to Immune Function Caused by UVR

Ultraviolet radiation, possibly through this capacity to elevate IL-1 production *in vivo*, also possesses the ability to cause a systemic depression in the capacity of exposed animals to develop cellular immune responses following antigenic challenge. Such systemic alterations can be distinguished from local UVR effects since these are limited to antigen presentation through the site of exposure. In systemic effects of UVR there is no concern for the exact site of antigen sensitization since UVR-exposed animals exhibiting systemic effects (e.g., acute-phase response) are systemically suppressed from being able to develop normal cellular immune responses.

It has been previously demonstrated that reductions in the intensity of the cellular immune response which accompany the consequences of UVR exposure are not paralleled by similar alterations in humoral immunity. Antibody production proceeds down a different pathway and requires a distinct pattern of lymphokines from cellular immunity to facilitate the activation, propagation and secretory potential of B cells.

At this juncture, we have only discussed a few cell types that are directly affected by UVR. There exists quite a list of other cell types within the skin that are equally involved in maintaining the full integrity of a normal immune response.

Many investigators who study skin immunology have recently come to the conclusion that the skin is an extremely important component of the mammalian immune system. Specific immunologic circuits have been described that function primarily within the confines of skin. The major immunologic circuit of skin has been given the acronym "SALT" for "skin-associated lymphoid tissue." Supporters of this concept believe in the existence of an integrated network between the skin and the various secondary lymphoid organs which drain any given skin site.

Interleukin-2 (IL-2) and interleukin-4 (IL-4) represent 2 lymphokines produced at various levels by antigen-activated T cells. The capacity to produce one of these two lymphokines is absolutely essential to immunologic responses controlled or mediated by T cells since they serve as T-cell growth factors needed for clonal expansion of antigen-specific T cells. IL-2 and IL-4 differ, however, in the influences they exert on other cell types. Interleukin-2, in addition to being a T-cell growth factor, is also necessary for the development of primary immunity, as well as in facilitating the differentiation of cellular immune responses. In contrast, interleukin-4 is capable of promoting B-cell proliferation, is a T-cell growth factor, and is also not appreciated to exert regulatory influences over γ-interferon (T-cell), tumor necrosis factor, interleukin-1, and prostaglandin (macrophage) production.

It is well documented that cellular immune function is systemically depressed in UVR-exposed animals, while the capacity to produce antibodies is not significantly altered. These historical findings suggested that UVR may be influencing the capacity of animals to produce either IL-2 or IL-4. Experimental evidence supported such a working hypothesis. Stimulation of spleen or lymph node T cells from normal mice with mitogens or specific antigens results in a significant rise in the supernatant titer of IL-2 and small titers of IL-4. When lymphoid organs from UVR-exposed donors are analyzed, the normal pattern of T-cell growth factor production is reversed, with IL-4 now representing the dominant species.

Gamma-interferon (γIFN) represents another essential lymphokine to the immune system since it possesses a diverse range of biological activities on many cell types. We now appreciate that γIFN also exhibits reduced production by T cells from UVR-exposed animals.

In an attempt to delineate the mechanism by which UVR affects the T-cell production of IL-2 and IL-4, we questioned the possible role played by elevations in interleukin-1 which results from UVR exposure. IL-1 levels are significantly elevated in UVR-exposed animals. Normal animals that are exposed to UVR, as well as animals exposed to a recombinant form of interleukin-1, were found to exhibit a shift in the pattern of T-cell lymphokines produced, with IL-2 production being diminished and IL-4 production augmented.

These effects of both UVR and IL-1 exposure persist for quite a long time following a single exposure or treatment. Further, if animals are continually exposed to UVR, or are maintained on low daily doses of IL-1, their IL-4 production dominates their IL-2 production indefinitely.

Role of Glucocorticosteroids in UVR Modifications to Immune Function

IL-1, UVR, and lipopolysaccharide (LPS) exposure of normal mice were each capable of shifting their capacity to produce IL-2 and IL-4 in a similar manner. Each of these inflammatory agents is also known to be capable of enhancing endogenous glucocorticosteroid (GCS) production through the capacity of IL-1 to stimulate both the pituitary and the hypothalamus. Linkage was established by determining that: (1) the administration of low-dose GCS to normal mice caused a decrease in IL-2 and a concomitant increase in IL-4 production, and (2) metyrapone (an 11β-hydroxylase inhibitor) and RU486 (a potent anti-glucocorticoid) were both able to abrogate the capacity of UVR, IL-1, and LPS to change lymphokine-secreting potential of isolated T cells from exposed animals. GCS are known to function as ligands for specific intracellular GCS receptors. These intracellular receptors can facilitate the transcriptional modulation of many specific genes. It appears, therefore, that GCS-receptor complexes possess the capacity to transcriptionally enhance the interleukin-4 gene, and simultaneously transcriptionally repress the production of the interleukin-2 gene, as well as the γIFN gene. These findings could account for many of the changes in immune function observed in UVR-exposed animals.

Conclusion

In summary, we now appreciate that UVR has a wide variety of effects on the mammalian immune system. Specifically, it is capable of exerting local effects that are mediated almost exclusively through the ability of this physical agent to functionally inactivate skin antigen-presenting cell function. This has the potential of reducing the natural barrier defense system at UVR-exposed sites, creating temporary windows where infectious agents could penetrate without immunological intervention.

There are also systemic modifications that are far more chronic than the local inactivation of skin-associated antigen-presenting cells. We now believe that many of these chronic alterations to immune function are facilitated through transient or chronic elevations in plasma GCS levels caused indirectly by UVR. These elevations in GCS are caused by the known inflammatory nature of UVR on epidermal keratinocytes, stimulating them to produce high levels of IL-1. This cytokine is a potent stimulant of both the pituitary and the hypothalamus, resulting in elevations in ACTH secretion and function. GCS have important regulatory influences on T-cell and macrophage function. Depressions in IL-2 and γIFN synthesis, and concomitant increases in the production of IL-4, could account for many of the immunologic changes that are observed in UVR-exposed animals.

References

Daynes, R.A., D. Kim Burnham, C.W. DeWitt, L.K. Roberts and G.G. Krueger. "The immunobiology of ultraviolet-radiation carcinogenesis." *Cancer Surveys*, 4:52 (1985).

Daynes, R.A. and J.D. Spikes. *Experimental and Clinical Photoimmunology*. Vol. I. Boca Raton, FL: CRC Press (1983).

Daynes, R.A. and G.G. Krueger. *Experimental and Clinical Photoimmunology*. Vols. II and III. Boca Raton, FL: CRC Press (1983 and 1986).

Kripke, M.L. "Immunological unresponsiveness induced by ultraviolet radiation." *Immunological Reviews*, 80:87 (1984).

Parrish, J.A., M.L. Kripke and W.L. Morison. *Photoimmunology*. New York, NY: Plenum Medical Book Company (1983).

EFFECTS OF UVB ON INFECTIOUS DISEASES

Suzanne Holmes Giannini

University of Maryland School of Medicine
Department of Medicine
Division of Geographic Medicine and
Center for Vaccine Development
Baltimore MD 21201

Abstract

Infectious diseases remain serious public health problems in the tropics and semi-tropics, which are exposed to high levels of UVB in sunlight. The UVB flux at the equator is one to four watts per square meter, so that even the highest experimentally effective doses can be acquired in several hours' exposure to midday sun. UVB is a potent suppressor of cell-mediated immunity. Many infectious diseases are acquired via the cutaneous route and protection is mediated by cellular immunity in immunocompetent individuals. In animal models for several such infectious diseases, UVB irradiation alters pathogenesis and immunity. In a murine model for cutaneous leishmaniasis, UVB irradiation applied during the first exposure to Leishmania altered the pathology of primary skin lesions, abrogated the induction of protective immunity, and may have predisposed the mice to fatal systemic leishmaniasis. UVB irradiation of mouse skin causes a spectrum of immunological and fine structural alterations also seen in human skin. Therefore, it is likely that exposure to high levels of solar UVB may enhance severity and adversely affect protective immune responses in naturally acquired infectious diseases of humans.

Physiological UVB Irradiances and the Immune Response

Most solar UV reaching the earth's surface is UVA (wavelengths between 320-400 nm). The shorter wavelength UVB (290-320 nm) is the more damaging biologically; its flux varies greatly depending on latitude, season and thickness of the ozone layer which absorbs these wavelengths [1]. UVC (180-290 nm) is completely blocked by the atmosphere.

UVB causes an impressive array of immunological effects, including depression of cell-mediated immune responses in contact hypersensitivity [2-8] and delayed type hypersensitivity [9-11], altered rates of secretion of cytokines [12,13; reviewed in 14], alteration in lymphocyte homing patterns [15,16], and the induction of T-suppressor cells [10,17,18] and of IL-4 secreting T-helper cells [19]. The effective doses of UVB range from 150 to 300 Joules (J) per square meter to damage epidermal antigen-presenting cells, the Langerhans cells [20], to 22 to 34 kiloJoules (kJ) m^{-2} to alter lymphocyte homing patterns [15], 49 kJ m^{-2} to cause systemic suppression of contact hypersensitivity [21] and 86 kJ m^{-2} to cause immunological unresponsiveness to UV-induced tumors [22].

In the latitudes between +25° North and –25° South, the midday UVB flux at the ground varies on clear days from 2.4 to 4.0 watts per square meter [calculated from ref. 1], so that over a 10-hour period, humans can be readily exposed to even the highest experimental level of biologically effective UVB, 86 kJ. Systemic immunosuppressive effects depend on the cumulative dose of UVB and not the rate of exposure [23], so that doses that are fractionated and delivered over several days are as effective as when they are given all at once. Furthermore, the effects of UVB-induced suppression of immunity persist for months [9,22]. Thus it is likely that the immunological effects of even kilojoule exposures of UVB may be physiologically relevant.

What are the effects of solar UVB irradiation on the body's response to infectious agents penetrating the skin? There are few experimental data to answer this critical question. Analysis is further complicated because infection and disease outcome are also affected by other factors, including the immunogenetic background of the host, the pathogenicity of the infecting organism, and environmental factors such as humidity, temperature, vector density and efficiency, concomitant infections and host nutritional state. The issue is very complex, but a growing body of evidence suggests that UVB irradiation can affect both pathogenesis and immunity in infectious diseases involving skin.

UVB and Infectious Diseases in Humans

The effect of UV irradiation on the progression of skin disease has been the subject of study for almost a century [reviewed in 24]. In 1880, the Danish physician Niels Finsen discovered that UV (which he called "chemical rays") was the component of sunlight that caused sunburn. In 1894, hypothesizing that UV promoted scarring in smallpox, he sequestered patients in rooms thickly hung with red curtains. They healed their pustules without scarring [25]. In 1901, he applied UV irradiation to treat lupus vulgaris, or tuberculosis of the skin [26], for which he was awarded the Nobel Prize in 1903.

"What is more natural," Finsen observed [25], "than that chemical rays should exert an injurious influence upon a diseased skin, when we see such severe inflammation produced by its influence upon the healthy skin?" Infectious diseases likely to be affected by solar UV are of two types: (1) Diseases of the skin itself and (2) diseases in which the skin is the portal of entry for the infectious agent, which manifests itself in other organs.

Infectious Diseases of Skin

Beneficial effects of UVB on human skin disease have been shown for three of these. Besides lupus vulgaris [26], which Finsen showed was cured by UV, we may add erysipelas, which was treated by actinic therapy until the advent of sulfa drugs [reviewed in 24]. In a more recent study, the skin lesions of herpes zoster were ameliorated by irradiation with erythemal doses of UVB [27]. It is important to note that, despite the seemingly beneficial effect of UV on the appearance of infected skin, no long-term follow-up was done in these studies with respect to numbers of viable organisms in irradiated skin or later incidence of systemic disease.

In four other infectious diseases of skin, UVB irradiation caused adverse effects in humans. Smallpox lesions are worsened by exposure to sunlight, as Finsen showed [25]. The lesions of herpes simplex virus type I (HSV-I) and type II (HSV-II) are reactivated by exposure to UVB [28-30]. Epidemiologic evidence suggests that exposure of immunosuppressed patients to sunlight leads to increased incidence of viral warts caused by papillomavirus [31,32], presumably due to solar UVB.

Other skin diseases which are likely to be affected by UVB irradiation include infections with the protozoan *Leishmania* species in Oriental sore or diffuse cutaneous leishmaniasis, with the nematode worm *Onchocerca volvulus* in onchocerciasis, with fungi such as Dermatophytes, *Candida albicans*, *Sporotrichum schenckii* and Ascomycetes, with *Mycobacterium leprae* in Hansen's disease, and with measles virus. The effect of UVB on these infections has not been established in humans, although some information is available from experimental animal models (discussed below).

Diseases In Which the Skin Is the Portal of Entry

The second group of infectious diseases likely to be affected by UVB includes those with a primary cutaneous phase, during which the organisms replicate and may elicit protective immunity to subsequent infections. For these agents, the skin-associated lymphoid tissue (SALT) is the first encounter with the host's immune system. For many of these diseases, protective immunity depends on the cellular arm of the immune response, which is altered by exposure of the SALT to UVB. These diseases include visceral and mucocutaneous leishmaniasis, schistosomiasis, hookworm, African sleeping sickness, Chagas' disease, yaws, cutaneous diphtheria, bubonic plague, anthrax, and perhaps Hansen's disease. Little is known about the effects of UVB irradiation on progression of any of these diseases in humans.

Because of the number of confounding variables, it is difficult to apply epidemiology to correlate levels of ambient UVB in a given geographic area with the prevalence of infectious diseases in the indigenous population. Even within the same geographic area, individual behavior (wearing hats, for example) markedly affects the dose of UVB striking the skin [33], resulting in large variation in person-to-person exposure levels. For these reasons the effects of UVB on infectious disease progression and immunity may be better studied in animal models.

UVB and Infectious Diseases in
Experimental Animal Models

In a variety of animal models, UVB irradiation affected pathology and/or immunity, although the effects of UVB may be influenced by timing of UVB irradiation relative to exposure to the infectious agent. UVB irradiation altered primary lesion development in cutaneous leishmaniasis [34-36], HSV-II [37,38] and *Moraxella bovis* [39]. UVB reactivated healed skin [40] and corneal [41] lesions caused by HSV. Increased numbers of organisms in UVB-irradiated animals were seen in infections with *Mycobacterium bovis* [42], *Plasmodium yoelii* and *P. chabaudi* [43]. Delayed-type hypersensitivity responses were abrogated in animals receiving UVB during the first exposure to *Leishmania major* [34,44], herpes simplex virus I [9,10] and II [37], *P. yoelii* [43], *Candida albicans* [46] and *Mycobacterium bovis* [42]. UVB irradiation applied to skin at the primary site of infection with *Leishmania major* depressed the protective immune response to subsequent infections [36,44]. UVB-irradiated mice infected with HSV-II [37], *L. major* [36] or *P. chabaudi* [43] developed high mortality rates in some experiments, compared with controls; these agents are usually not lethal.

The effects of measured doses of UVB on infectious diseases in experimental mouse models are summarized in Table I. It is evident from these observations that UVB profoundly affects the ability of the host to mount an immune response against invading microorganisms. This is well illustrated by a mouse model system for leishmaniasis.

Features of Leishmaniasis in Humans

Leishmaniasis is an excellent model to study the effects of UVB on infectious diseases. The causative agent is deposited in exposed skin by crepuscular sandfly vectors, themselves not exposed to UVB. Leishmania grow in the upper layer of the dermis or the epidermis, which also may be exposed to UVB in sunlight. Disease progresses in two phases: (1) replication of organisms in the skin, where they may or may not induce an ulcer (Oriental sore, the most common form of leishmaniasis); and (2) in some individuals, dissemination of the organisms beyond the primary skin site and draining lymph nodes to distal skin (to cause incurable diffuse cutaneous leishmaniasis), mucocutaneous membranes (mutilating mucocutaneous leishmaniasis, or espundia) or visceral lymphoid organs (systemic, often fatal, visceral leishmaniasis or kala-azar). Leishmaniasis is endemic on every continent except Australia and the Antarctic [47], so that epidemiological studies could be designed in the future to evaluate UVB exposure as a risk factor in disease development. Other known risk factors are the species of parasite [47] and the immunogenetic background of the host [48,49].

Effect of UVB on Murine Cutaneous Leishmaniasis

In B10.129(10M)Sn [34,35] and in C3H/HeJ mice [36], suberythematous doses of UVB irradiation alter development of skin lesions during the primary phase of leishmaniasis. Although having a seemingly beneficial effect on skin pathology, this low dose of UVB leaves the irradiated mice nonimmune [36,44]. It is important to note that in this model the lack of development of overt cutaneous disease after UVB irradiation is accompanied by two negative factors: (1) loss of a diagnostic indicator that infection has occurred; and (2) despite the absence of lesions, UVB-exposed infected individuals remain infected systemically. At a later time, immunosuppression caused by environmental or other factors could reactivate cryptic Leishmania to cause systemic disease, such as visceral or diffuse cutaneous leishmaniases. Such reactivation is well documented in patients, cryptically infected with Leishmania, who later develop acquired immunodeficiency syndrome (AIDS) [50,51].

Development of skin lesions and immunity in murine cutaneous leishmaniasis is altered by the lowest doses of UVB that are active in classical photoimmunological studies [34,36,44]. That suberythematous levels of UVB are active suggests that the model system is a sensitive indicator for immunomodulatory effects of UVB. Similar doses of UVB also affect resistance to herpes simplex viruses and murine malaria [9,10,37,43].

Irradiation with broadband UVB (290-320 nm) alters the development of the primary phase of disease in UVB-sensitive mouse strains, but does not reduce parasite numbers. Irradiances of UVB that suppressed the induction of contact hypersensitivity to dinitrofluorobenzene (600 J m^{-2}), abrogated the development of primary skin lesions in male C3H/HeJ mice infected with 10^6 L. major. Superficially UVB irradiation appeared to be beneficial, but upon closer inspection we find it was not. UVB-irradiated and nonirradiated mice were equally heavily parasitized, both at the injection site and at the draining lymph nodes (data not shown). Similar findings were seen in the B10.129(10M)Sn model [34].

These observations suggest that UVB does not reduce viability of parasites in skin. They also show that skin pathology is a poor indicator of parasite numbers or the extent of parasite dissemination, the second phase of leishmanial disease. In subtropical and tropical regions endemic for leishmaniasis, surveys that are limited to inspection of skin lesions may be grossly inadequate to detect infected individuals and estimate prevalence of leishmaniasis.

Broadband UVB irradiation of Leishmania-infected skin does not alter pathogenicity of parasites. Amelioration of primary skin lesions by UVB applied to the injection site might have been caused by nonlethal changes in parasite pathogenicity. To test this, Leishmania were cultivated from skins of UVB-irradiated and nonirradiated mice from three separate experiments. Leishmania were injected into normal mice and lesions were measured to estimate parasite pathogenicity. No significant differences were observed between the pathogenicity of organisms recovered from UV-irradiated or control mice.

Broadband UVB applied to the primary injection site abrogates protective immunity against a subsequent reinfection at a distal skin site. Infection with *L. major* generally confers immunity to subsequent infections, so that skin lesions resulting from a later challenge infection are reduced in size or do not develop. UVB-irradiated or non-irradiated C3H/HeJ mice were infected with *L. major* and then reinfected six weeks later in a nonirradiated distal skin site. Protective immunity was evident in the nonirradiated mice, in which lesions did not develop at the distal challenge site, nor were any Leishmania detected. But protective immunity was lacking in the mice whose first exposure to Leishmania was via UVB-irradiated skin; they developed nodules at the challenge site, from which parasites were recovered (data not shown).

UVB irradiation may convert cutaneous leishmaniasis from a nonlethal to a lethal disease. In three studies of the effects of narrow band UVB on cutaneous leishmaniasis in C3H/HeJ mice, the mortality rates for *L. major*-infected mice which had also been UVB irradiated were substantially greater (a total of ~30% dead) than for nonir-radiated, *L. major*-infected mice (0% dead), or for noninfected, UVB-irradiated mice (0% dead). These findings were made retrospectively from three different experiments in which observation time to necropsy varied from one to four months; the total exposures of UV at 290 nm ranged from 1000 to 1675 J m^{-2}; the doses of UV at 320 nm ranged from 1000 to 7800 J m^{-2}.

Because of the variability in exposures and times of observation, it is difficult to interpret these preliminary observations quantitatively, but they clearly indicate that further study of the effects of UVB on mortality in leishmaniasis is warranted. The significance of increased mortality is that systemic visceral leishmaniasis often occurs in fulminating epidemics in the absence of concomitant increases in numbers of reservoir hosts or vectors. Visceral leishmaniasis frequently is not preceded by a skin lesion (although it may be). This picture in humans resembles that seen in our mice, where effects of UVB on skin pathology and parasite dissemination to internal organs proceed independently.

Conclusions

Sunlight in the tropics and semitropics currently contains levels of UVB sufficient to deliver a dose that can be immunosuppressive in several experimental immunological systems, with effects ranging from local suppression of contact hypersensitivity to systemic suppression of tumor immunity. In these systems, the effects do not depend on the rate of exposure, but rather the total dose, i.e. the effects are cumulative. The levels of UVB reaching the earth's surface are inversely proportional to the thickness of the ozone layer, so that ozone layer depletion will likely increase UVB levels in temperate and arctic zones as well as in the tropics and semitropics.

Epidemiologic studies in humans suggest that exposure to sunlight (presumably the UVB component) may alter pathology and susceptibility to some infectious diseases. Exposures to levels of UVB, attainable in the tropics and semitropics, can affect

pathogenesis and immunity in animal models for several infectious diseases. Two important features of UVB exposure on infectious disease development have been determined
in a mouse model for cutaneous leishmaniasis: (1) loss of a diagnostic indicator of infection; and (2) despite the absence of visible skin lesions, infected individuals remain infected systemically. Further research is needed to determine for UVB the effective
exposures, action spectra and cellular targets modifying pathogenesis and protective immunity in these disease models.

TABLE I. EFFECTS OF UVB IRRADIATION ON RESPONSE TO INFECTIOUS AGENTS IN EXPERIMENTAL ANIMAL MODELS

[REFERENCE CITED]

DISEASE MODEL:			EFFECTS OF UVB IRRADIATION ON HOST RESPONSES:				
Infectious agent	Mouse strain	Dose of UVB (kJ)*	Depressed DTH*	Altered pathology	Increased no. orgs.	Depressed protection	Enhanced mortality
Leishmania major	B10.129-(10M)Sn	2	[34]	[34,35]		[44]	
	C3H/HeJ	8		[36]		[36]	[36]
Herpes simplex I	C3H/f/-bu/Kam	1	[9,10]				
Herpes simplex II	BALB/c	1 – 3	[37]	[37]			[37]
Candida albicans	C3H/HeN-Cr(MTV-)	47	[46]				
Mycobacter-ium bovis (BCG)	BALB/c-AnNCr	40 – 45	[42]	[42]+	[42]		
	C3H/HeN-Cr(MTV-)	40 – 45	[42]	[42]	[42]		
Plasmodium yoelii	BALB/c	3	[43]		[43]		
Plasmodium chabaudi	BALB/c	3			[43]		[43]

* Except for [44] all studies used fluorescent sunlamps with ~60% of output in the UVB range. Published doses have been rounded off to the nearest kJ m^{-2}.
+ Size of draining lymph node; in all other models, skin lesions were altered.

Acknowledgments

Supported in part by grants from the Medical Biotechnology Center of the Maryland Biotechnology Institute. Dr. Edmond A. Goidl (University of Maryland School of Medicine) and Dr. Janice Longstreth (Clement Associates) provided helpful comments and discussion.

References

1. Frederick, J.E. 1986. "The ultraviolet radiation environment of the biosphere". *In* J.G. Titus [ed.], *Effects of Changes in Stratospheric Ozone and Global Climate, Vol. I. Overview.* Washington, D.C.: U.S. Environmental Protection Agency, pp. 121–128.

2. Noonan, F.P., E.C. De Fabo and M.L. Kripke. 1981. "Suppression of contact hypersensitivity by UV radiation and its relationship to UV-induced suppression of tumor immunity." *Photochem. Photobiol.* 34:683–689.

3. Noonan, F.P., M.L. Kripke, G.M. Pedersen and M.I. Green. 1981. "Suppression of contact hypersensitivity in mice by ultraviolet irradiation is associated with defective antigen presentation." *Immunology* 43:527–533.

4. De Fabo, E.C. and F.P. Noonan. 1983. "Mechanism of immune suppression by ultraviolet irradiation *in vivo*. I. Evidence for the existence of a unique photoreceptor in skin and its role in photoimmunology." *J. Exp. Med.* 157:84–98.

5. Toews, G.B., P.R. Bergstresser and J.W. Streilein. 1980. "Epidermal Langerhans cell density determines whether contact hypersensitivity or unresponsiveness follows skin painting with DNFB." *J. Immunol.* 124:445–453.

6. Noonan, F.P., C. Bucana, D.N. Sauder and E.C. De Fabo. 1984. "Mechanism of systemic immune suppression by UV irradiation *in vivo*. II. The UV effects on number and morphology of epidermal Langerhans cells and the UV-induced suppression of contact hypersensitivity have different wavelength dependencies." *J. Immunol.* 132:2408–2416.

7. Kripke, M.L. and E. McClendon. 1986. "Studies on the role of antigen-presenting cells in the systemic suppression of contact hypersensitivity by UVB irradiation." *J. Immunol.* 137:443–447.

8. Morison, W.L., C. Bucana and M.L. Kripke. 1984. "Systemic suppression of contact hypersensitivity by UVB irradiation is unrelated to the UVB induced alterations in the morphology and number of Langerhans cells." *Immunology* 52:299–306.

9. Howie, S., M. Norval and J. Maingay. 1986. "Exposure to low-dose ultraviolet radiation suppresses delayed-type hypersensitivity to herpes simplex virus in mice." *J. Invest. Dermatol.* 86:125–128.

10. Howie, S.E.M., M. Norval, J. Maingay and J.A. Ross. 1986. "Two phenotypically distinct T cells ($Ly1^+2^-$ and $Ly1^{-2+}$) are involved in ultraviolet-B light-induced suppression of the efferent DTH response to HSV-1 *in vivo*." *Immunology* 58:653–658.

11. Hayashi, Y. and L. Aurelian. 1986. "Immunity to herpes simplex virus type 2: Viral antigen-presenting capacity of epidermal cells and its impairment by ultraviolet irradiation." *J. Immunol.* 136:1087–1092.

12. Ansel, J., T.A. Luger, A. Kock, D. Hochstein and I. Green. 1984. "The effect of *in vitro* irradiation on the production of IL 1 by murine macrophages and P388D$_1$ cells." *J. Immunol.* 133:1350–1355.

13. Gahring, L., M. Baltz, M.B. Pepys and R. Daynes. 1984. "Effect of ultraviolet radiation on production of epidermal cell thymocyte activating factor/interleukin 1 *in vivo* and *in vitro*." *Proc. Natl. Acad. Sci.* (USA) 81:1198–1202.

14. Luger, T.A. 1986. "UVL and epidermal cell cytokine production." *Photodermatology* 3:123–124.

15. Spangrude, G.J., E.J. Bernhard, R.S. Ajioka and R.A. Daynes. 1983. "Alterations in lymphocyte homing patterns within mice exposed to ultraviolet radiation." *J. Immunol.* 130:2974–2981.

16. Daynes, R.A., G.J. Spangrude, L.K. Roberts and G.G. Krueger. 1985. "Regulation by the skin of lymphoid cell recirculation and localization properties." *J. Invest. Dermatol.* 85:14s–20s.

17. Fisher, M.S. and M.L. Kripke. 1982. "Systemic alteration induced in mice by ultraviolet light irradiation and its relationship to ultraviolet carcinogenesis." *Proc. Nat. Acad. Sci.* (USA) 74:1688–1692.

18. Granstein, R.D. 1985. "Epidermal I-J bearing cells are responsible for transferable suppressor cell generation after immunization of mice with ultraviolet radiation-treated epidermal cells." *J. Invest. Dermatol.* 84:206–209.

19. Araneo, B.A., T. Dowell, H.B. Moon and R.A. Daynes. 1989. "Regulation of murine lymphokine production *in vivo*. Ultraviolet radiation exposure depresses IL-2 and enhances IL-4 production by T cells through an IL-1-dependent mechanism." *J. Immunol.* 143:1737–1744.

20. Aberer, W., G. Schuler, G. Stingl, H. Honigsmann and K. Wolff. 1981. "Ultraviolet light depletes surface markers of Langerhans cells." *J. Invest. Dermatol.* 76:202.

21. Kripke, M.L. and W.L. Morison. 1986. "Studies on the mechanism of systemic suppression of contact hypersensitivity by ultraviolet B irradiation." *Photodermatology* 3:4–14.

22. De Fabo, E.C. and M.L. Kripke. 1979. "Dose-response characteristics of immunologic unresponsiveness to UV-induced tumors produced by UV irradiation of mice." *Photochem. Photobiol.* 30:385–390.

23. De Fabo, E.C. and M.L. Kripke. 1980. "Wavelength dependence and dose-rate independence of UV radiation-induced immunologic unresponsiveness of mice to a UV-induced fibrosarcoma." *Photochem. Photobiol.* 32:183–188.

24. Licht, S. 1983. "History of ultraviolet therapy." *In*: G.K. Stillwell (ed.), *Therapeutic Electricity and Ultraviolet Radiation*, 3rd Ed. Baltimore: Williams and Wilkins, pp. 174–193.

25. Finsen, N.R. 1901. "The chemical rays of light and smallpox." *In*: *Phototherapy*. London: Edward Arnold, pp. 1–36.

26. Finsen, N.R. 1901. "The treatment of lupus vulgaris by concentrated chemical rays." *In*: *Phototherapy*. London: Edward Arnold, pp. 73–75.

27. Szigeti, B., J. Chapiro and J.M. Saint-Girons. 1976. "Traitement du zona par actinotherapie ultraviolette." *J. Radiol. Electrol.* **57**:549–551.

28. Wheeler, C.E. Jr. 1975. "Pathogenesis of recurrent herpes simplex infections." *J. Invest. Dermatol.* **65**:341–346.

29. Spruance, S.L. 1985. "Pathogenesis of herpes simplex labialis: Experimental induction of lesions with UV light." *J. Clin. Microbiol.* **22**:366–368.

30. Klein, K.L. and C.C. Linnemann, Jr. 1986. "Induction of recurrent genital herpes simplex virus type 2 infection by ultraviolet light." *Lancet* **1**:796–797.

31. Boyle, J., R.M. MacKie, J.D. Briggs, B.J.R. Junor and T.C. Aitchison. 1984. "Cancer, warts and sunshine in renal transplant patients: A case-control study." *Lancet* **1**:702–705.

32. Dyall-Smith, D. and G. Varigos. 1985. "The malignant potential of papillomavirus." *Austral. J. Dermatol.* **26**:102–107.

33. Urbach, F. 1969. "Geographic pathology of skin cancer." *In*: F. Urbach (ed.), *The Biologic Effects of Ultraviolet Radiation*. New York: Pergamon Press, pp. 635–650.

34. Giannini, M.S.H. 1986. "Suppression of pathogenesis in cutaneous leishmaniasis by UV irradiation." *Infect. Immun.* **51**:838–843.

35. Giannini, S.H. and E.C. De Fabo. 1989. "Abrogation of skin lesions in cutaneous leishmaniasis by ultraviolet B irradiation." *In*: D.T. Hart (ed.), *Leishmaniasis: The Current Status and New Strategies for Control*. London: Plenum Publishing Corp., pp. 677–684.

36. Giannini, S.H. "Effects of UVB irradiation on pathogenesis and immunity in the C3H/HeJ mouse." Ms. in preparation.

37. Yasumoto, S., Y. Hayashi and L. Aurelian. 1987. "Immunity to herpes simplex virus type 2: Suppression of virus-induced immune responses in ultraviolet B-irradiated mice." *J. Immunol.* **139**:2788–2793.

38. Harbour, D.A., T.J. Hill and W.A. Blyth. 1977. "The effect of ultraviolet light on primary herpes simplex virus infection in the mouse." *Arch. Virol.* **54**:367–372.

39. Gerber, J.D. and S.K. Frank. 1983. "Enhancement of *Moraxella bovis*-induced keratitis of mice by exposure of the eye to ultraviolet radiation and ragweed extract." *Amer. J. Vet. Res.* **44**:1382–1384.

40. Blyth, W.A., T.J. Hill, H.J. Field and D.A. Harbour. 1976. "Reactivation of herpes simplex virus infection by ultraviolet light and possible involvement of prostaglandins." *J. Gen. Virol.* **33**:547–550.

41. Spurney, R.V. and M.S. Rosenthal. 1972. "Ultraviolet-induced recurrent herpes simplex virus keratitis." *Amer. J. Ophthalmol.* **73**:609–610.

42. Jeevan, A. and M.L. Kripke. 1989. Effect of a single exposure to ultraviolet radiation on *Mycobacterium bovis* bacillus Calmette-Guerin infection in mice." *J. Immunol.* **143**:2837–2843.

43. Taylor, D.W. and D.A. Eagles. 1989. Assessing the effects of ultraviolet radiation on malarial immunity. Submitted to Sabotka and Company in fulfillment of Subcontract No. 132.914, EPA Contract No. 68–01–7288.

44. Giannini, S.H. 1986. "Effects of UVB on infectious disease." *In*: J.G. Titus (ed.), *Effects of Changes in Stratospheric Ozone and Global Climate. Volume 2: Stratospheric Ozone*. Washington, D.C.: U.S. Environmental Protection Agency, pp. 101–112.

45. Aurelian, L., S. Yasumoto and C.C. Smith. 1988. "Antigen-specific immune-suppressor factor in herpes simplex virus type 2 infections of UV B-irradiated mice." *J. Virol.* **62**:2520–2524.

46. Denkins, Y., I.J. Fidler and M.L. Kripke. 1989. "Exposure of mice to UVB radiation suppresses delayed hypersensitivity to *Candida albicans*." *Photochem. Photobiol.* **49**:615–619.

47. Anonymous. 1984. "The Leishmaniases." *WHO Tech. Rep. Ser.* **701**.

48. Greenblatt, C.L. 1980. "The present and future of vaccination for cutaneous leishmaniasis." *Prog. Clin. Biolog. Res.* **47**:259–285.

49. Walton, B.C. and L. Valverde. 1979. "Racial differences in espundia." *Ann. Trop. Med. Parasitol.* **73**:23–29.

50. Fernandez-Guerrero, M.L., J.M. Aguado, L. Buzon, C. Barros, C. Montalban, T. Martin and E. Bouza. 1987. "Visceral leishmaniasis in immunocompromised hosts." *Amer. J. Med.* **83**:1098–1102.

51. Clauvel, J.P., L.J. Couderc, J. Belmin, M.T. Daniel, C. Rabian and M. Seligman. 1986. "Visceral leishmaniasis complicating acquired immunodeficiency syndrome." *Trans. Roy. Soc. Trop. Med. Hyg.* **80**:1011–1012.

INFECTIOUS DISEASES AND ATMOSPHERIC CHANGE

Robert E. Shope

Department of Epidemiology and Public Health
Yale University School of Medicine
Box 3333
New Haven, CT 06510

Abstract

The geographic limits of arthropod-borne and other zoonotic diseases are determined primarily by rainfall and temperature. Certain infectious agents vectored by mosquitoes adapt readily to movement with humans and domestic animals; these are most likely to relocate with warming of the temperate and arctic zones. These include dengue and yellow fever viruses, transmitted by *Aedes aegypti*, and St. Louis encephalitis virus, transmitted in epidemic form by *Culex pipiens* complex mosquitoes. Major U.S. population centers now too cold to support these agents may become epidemic centers. Leishmaniasis is a protozoon transmitted by phlebotomine sand flies; some New World sand fly species are very adaptable with climate change and may establish themselves as vectors of *Leishmania* in the southern United States if warming occurs. Warming may also shift northward the range of the vampire bat, *Desmodus rotundus*, a major reservoir of rabies virus. Once established in Texas and other parts of the southern United States, this bat could transmit rabies to cattle and humans. More intensive study of the climatic factors that limit the distribution of these agents and their vectors would permit us to predict the consequences of climate change and to intervene in the transmission cycles.

Introduction

Most human infectious diseases are not directly influenced by the climate. Nor are the experts in accord as to "if, when or how much" the climate will change (1). Therefore this will be a discussion of what might happen. I shall let the reader decide the "if, when and how much" of atmospheric change.

The principles that will determine which of the infectious diseases will become more prevalent in North America with global warming are straightforward. The ecology of the disease affected will have some or all of these attributes:

1. The distribution of the disease is focal.

2. The focality depends primarily on the range of a nonhuman reservoir, i.e. animal, arthropod, plant, soil or substance, or a combination of these.

3. The range of the reservoir depends on temperature, water level, and/or rainfall. If the agent is vectored by an arthropod, the arthropod will develop more rapidly, and the agent will replicate more rapidly or efficiently at warm temperature.

4. The reservoir and the agent will survive atmospheric change.

5. The reservoir is capable of translocation to a new geographic area if atmospheric change ensues.

6. The reservoir is capable of adaptation to the new geographic area.

Some diseases of human beings such as measles and poliomyelitis are not focal and their distribution does not depend on a nonhuman reservoir. It is unlikely that the prevalence or severity of measles or poliomyelitis will be changed if global warming occurs. However, these diseases are modified by socioeconomic conditions and, to the extent that global atmospheric change also impoverishes a region, diseases such as measles and poliomyelitis may become more prevalent.

Examples of Infectious Diseases Modified by Global Atmospheric Change

1. St. Louis Encephalitis

Some infectious agents that are vectored by mosquitoes adapt readily to movement with human beings; these are the most likely to relocate with warming of the temperate and arctic zones. Among these diseases are dengue and yellow fever, the causative viruses being transmitted by *Aedes aegypti* mosquitoes, and St. Louis encephalitis, the causative virus being transmitted in epidemic form by *Culex pipiens* complex mosquitoes. St. Louis encephalitis will be used to illustrate how change may occur in the case of mosquito-borne virus diseases.

St. Louis encephalitis virus causes large outbreaks of encephalitis in the United States about every 10 years. During intervening years, the infection is endemic with smaller outbreaks and a few sporadic cases in the western United States, especially in Texas, Colorado and California. The virus is limited to the Americas. This disease is unusually severe in elderly individuals with case fatality during epidemics of between 20% and 80% in those over 60 years of age (2).

St. Louis encephalitis has a rural or enzootic cycle involving *Culex tarsalis* mosquitoes and birds. *Culex tarsalis* is a ground-pool breeder of the western U.S., prevalent especially in irrigated farm lands. The urban or epidemic cycle of St. Louis encephalitis involves birds and *Culex quinquefasciatus* in the southern U.S. or the closely-related *Culex pipiens* mosquito in the North. These mosquitoes require water that is rich in organic matter, and they thrive in sewage-containing water. The effect of temperature and rainfall on epidemics of St. Louis encephalitis is well documented and has been reviewed by Monath (3).

Epidemics usually occur south of the 70-degree F June isotherm. This isotherm can be expected gradually to move northward during global warming, and epidemics can also be expected more frequently in northern U.S. cities. Exceptionally there have been epidemics north of the 70-degree F June isotherm, but only in especially warm years (4). Monath (3) examined temperature records from 15 sites during epidemic and non-epidemic years. The more northerly sites deviated the most from mean temperature in epidemic years. Warmer than normal winters and summers correlated with epidemics, but the April temperatures in epidemic years were consistently colder. Not only does temperature limit the range of the vector, but also the larvae develop more rapidly in warmer temperatures, and the incubation period of the virus in the mosquito is shorter in warmer climates.

Monath also examined precipitation (3) during 15 epidemic years and compared it with nonepidemic years. Precipitation was above normal in January and February and paradoxically low in July. In keeping with Monath's observations, rainfall in St. Louis in the summer of 1933 at the time of the initially recognized epidemic was the lowest since records had been kept (5). Low rainfall favors breeding of *Culex pipiens* mosquitoes since sewage-containing breeding sites are not washed out.

To summarize the evidence that atmospheric change will affect epidemic St. Louis encephalitis prevalence: (1) the disease is focal, (2) its focality depends on the vector mosquitoes of the *Culex pipiens* complex, (3) the range of the vector, the rapidity of development of the vector, and the rate of replication of the virus are limited by temperature. In addition, it is expected that (4) the reservoir (birds and *Culex pipiens*) and the virus will survive atmospheric change, (5) the vector will translocate northward if warming occurs, and (6) the vector will adapt well to ecological conditions in cities further north.

2. Phlebotomine Sand Flies and Leishmaniasis

Many parasitic diseases will probably not be greatly affected by global atmospheric change in their prevalence or severity in the United States. Leishmaniasis, however, may be an exception. *Leishmania chagasi* causes visceral leishmaniasis in the New World. The protozoal parasite is transmitted by the sand fly, *Lutzomyia longipalpis*. The reservoir is the domestic dog, *Canis familiaris*.

Visceral leishmaniasis, also known as Kala Azar, occurs in the New World primarily as a disease of young children. The children develop fever, enlarged spleen, loss of appetite, anemia, and depression of cellular immunity. Fatalities are often linked to secondary bacterial infections.

The distribution of *Lutzomyia longipalpis* matches very closely that of visceral leish-maniasis (6). Cases extend throughout the tropics from northern Argentina to Mexico. The sand fly has adapted to a peridomestic ecology, resting in chicken houses, sheds and corrals. It feeds on dogs. In Rio de Janeiro and cities in the Northeast of Brazil, the vector has become urbanized. Therefore, there is good reason to believe that with warmer temperatures the sand fly, the parasite, and the disease could become established initially in northern Mexico and later in Texas and other parts of the southern United States.

Visceral leishmaniasis caused by *Leishmania chagasi* has many of the same attributes as St. Louis encephalitis. The disease is focal and the distribution depends on the range of the sand fly vector. This range is limited by temperature. The vector, *Lutzomyia longipalpis* is highly adaptable, will likely survive atmospheric change, and may well translocate further north.

3. Vampire Bats and Rabies

Let us turn now to another example of a disease restricted in its distribution by ecological factors. Vampire bat-transmitted rabies has occurred for centuries, but has been recognized as a specific disease in the tropical Americas since the period of World War I when 10,000 cattle died south of Sao Paulo, Brazil (7). Later in Trinidad, Hurst and Pawan (8) established that the disease affected people as well as cattle. There, 53 people died during the early 1930s following bites of vampire bats (*Desmodus rotundus*) (9).

The vampire bat is a communal creature (10). These animals are infected with rabies virus presumably by the contaminated saliva of other bats during grooming, or by bites during fighting to maintain territorial rights. The vampire bat feeds upon the blood of animals and prefers in decreasing order of priority cattle, horses, chickens, dogs and human beings (11). The bats require blood for sustenance and are the only vertebrate animals known to feed entirely on blood (12).

At the present time, the vampire bat is restricted in its distribution to the tropics (13). The range is from central Argentina to northern Mexico. The restriction is presumably one of temperature, since there is ample food for vampire bats north of Mexico, especially in California and Texas. Texas is one of the largest cattle-raising areas of the world with about 13 million head; the cattle industry grosses in excess of $1.25 billion annually.

It has been warm enough in 1989 in Texas for cattle to graze throughout the year. If, through global atmospheric change, the temperature should rise sufficiently to support populations of vampire bats, predictably the bat populations would be large because of the immense food supply.

It is assumed that, with gradual global atmospheric change, the movement northward of vampire bats would also be gradual. The situation might be similar to the spread of bovine paralytic rabies in 1953 from southern Bolivia, where there were 260,000 cattle deaths, to northern Argentina over a period of 15 years. Eventually, large areas in Argentina were left with no cattle at all (14).

Acha estimated the annual mortality of cattle from bat-transmitted rabies about twenty years ago to be 514,500 animals worth $47.6 million (13). The numbers of human rabies cases are not known but are probably very small because the bat feeds at night on individuals sleeping out of doors. Those at risk, therefore, are persons involved in outdoor recreation, such as camping.

Should vampire bat-transmitted rabies become a problem in the United States, the costs of combatting it would not be insurmountable. Some Mexican ranchers successfully vaccinate their cattle today, and the U.S. ranchers could do likewise. The vampire bat is also susceptible to anticoagulants which are an effective means of control. The possible small numbers of human rabies cases in states such as California and Texas would make headlines but, to put it in perspective, would be much less significant than the current preventable mortality from lung cancer, accidents and AIDS.

Human Population Migration and Emerging Diseases

It is predicted that warming trends will be more marked in the temperate and arctic regions. One possible consequence is the gradual migration of human populations to these zones. In North America this could mean denser settlement in Canada and Alaska.

Infectious agents are present in all parts of the globe. In relatively isolated areas these agents have wildlife cycles that may involve the few local people, but immunize at an early age, and are not recognized as serious disease causing agents. Humans who settle and colonize these areas will be predictably nonimmune to the indigenous infectious agents and thus will represent virgin populations, subject to infections with any and all of the local agents. This situation is analogous to the colonization of Africa by Europeans in the 1800s when the colonists frequently succumbed to malaria and other infectious diseases.

In addition, future colonization of northern Europe and Asia, Canada and Alaska may involve settlements that are in today's wilderness areas. Infectious agents will be encountered that today do not infect people, simply because people are not there to be exposed. These will be zoonotic diseases that have cycles in wildlife and will newly emerge as causes of human disease.

How to Prepare for Global Atmospheric Change

We have much to learn about the effects of temperature and rainfall on the natural cycles of infectious agents. It is time now to study the agents of disease such as those named above. Their ability to adapt to changing conditions can best be explored at the geographic fringes of their range. They and their vertebrate hosts and vectors can be studied also in the laboratory where changes in temperature and rainfall can be simulated. We should be able to identify the known agents that meet the criteria for translocation in the event of global warming, and then devise methods for their control.

It is also timely to initiate studies of wildlife ecology and the infectious agents in Canada and Alaska. It is better to know the risks before human migration than after. With knowledge we may be able to prevent serious disease.

References

1. Abelson, P.H. "Uncertainties about global warming." *Science* **247**:1529, 1990.

2. Monath, T.P. *St. Louis Encephalitis*, chapter 6, p. 266. American Public Health Association, Washington, DC, 1980.

3. Monath, T.P. *St. Louis Encephalitis*, chapter 6, pp. 289–293. American Public Health Association, Washington, DC, 1980.

4. Hess, A.D., Cherubin, C.E., and LaMotte, L.C. "Relation of temperature to activity of western and St. Louis encephalitis viruses." *Am. J. Trop. Med. & Hyg.* **12**:657–667, 1963.

5. *Report of the St. Louis outbreak of encephalitis*. Public Health Bulletin No. 214, U.S. Government Printing Office, Washington, DC, 1935.

6. Grimaldi, G., Jr., Tesh, R.B., and McMahon-Pratt, D. "A review of the geographic distribution and epidemiology of leishmaniasis in the new world." *Am. J. Trop. Med. & Hyg.* **41**:687–725, 1989.

7. Haupt, H. and Rehaag, H. "Durch Fledermause verbreitete seuchenhafte Tollwut unter Viehbestanden in Santa Catharina (Sud-Brasielien)." *Z. Infektionskr. Hyg. Haustiere* **22**:104–127, 1921.

8. Hurst, E.W. and Pawan, J.L. "An outbreak of rabies in Trinidad without history of bites and with the symptoms of acute ascending myelitis." *Lancet II*, 622–628, 1931.

9. Hurst, 1936, quoted in Baer, G.M. "Bovine paralytic rabies and rabies in the vampire bat," chapter 10, pp. 156–157, vol. 2, *The Natural History of Rabies*, G.M. Baer ed. Academic Press, New York, 1975.

10. Arellano-Sota, C. "Biology, ecology, and control of the vampire bat." *Reviews of Infectious Diseases*, vol. 10, supplement 4, S615–S619, 1988.

11. Goodwin, G.G. and Greenhall, A.M. *Bull. Amer. Mus. Natur. Hist.* **122**:187–301, 1961.

12. Sulkin, S.E. and Allen, R. "Virus Infections of Bats," p. 5, *Monographs in Virology*, vol. 8, J.L. Melnick ed. S. Karger, Basel, 1974.

13. Acha, P.N. "Epidemiology of paralytic bovine rabies and bat rabies." *Bull. Off. Int. Epiz.* **67**:343–382, 1967.

14. Lopez Adaros, H., Silva, M., and La Mata, M. *Rabia paralitica en el norte Argentino*. "Seminario Sobre Rabia para Los Paises de la Zona IV, Bolivia, Colombia, Ecuador, Peru," pp. 161–203. Pan Amer. Health Organ., Buenos Aires, Argentina, 1969.

HUMAN NUTRITION AND ATMOSPHERIC CHANGE

Alexander Leaf

Department of Preventive Medicine
Harvard Medical School
Boston, MA

We have been hearing and will be hearing about the consequences of the direct effects of atmospheric changes on human health. But the most serious consequences now appear to be the late indirect effects, namely food shortages. In 1989, for the third year in a row, world grain production fell short of consumption. After increasing a phenomenal 2.6-fold from 1950 to 1984, world output has essentially leveled off while population continues to leap ahead, as L.R. Brown points out (1). The 1989 grain harvest of 1.66 billion tons is exactly the same as that of 1984, only there are 440 million more people to feed. This short-fall is experienced predominantly in the developing world. In 1980, there were 340 million people in 87 developing countries not getting enough calories to prevent stunted growth and serious health risks. This represents a 14% increase in numbers over the prior decade, and the World Bank predicts that these numbers are likely to continue to grow (2). Between the 1984 peak of production and 1988, world grain production per person fell 14%, while in Africa, since 1967, the decline has been 27% (3). These figures are but the harbinger of what the future portends.

I will try to sketch the consequences of the global environmental changes which are likely to play a prominent role in increasing food shortages and starvation in the future. But first I must deal with the root cause of all the problems we are discussing, the population explosion. Figure 1 shows what has been happening to the world's population. It took until 1850 before there were one billion humans on earth. Now we are adding another billion with each decade and the rate increases. Predictions by the United Nations (UN) are that this will continue until the population stabilizes at 8 to 14 billion sometime in the 21st century (4). Furthermore, it is expected that 90 percent of the increase will take place in the developing countries, and 90 percent of that will occur in cities that are already bursting. It is the needs and aspirations of the growing world population for a higher standard of living that motivates the accelerating consumption of the earth's resources and which has accelerated the greenhouse effect and pollution of the environment. I will not discuss the central population issue further, but it is important that we keep it in mind as we consider the effect that the global climate and environmental changes will have on the adequacy of food supplies in the future.

The predicted mean global warming from the accumulation of greenhouse gases in the atmosphere of 2 to 5°C (5) potentially can be disruptive to agriculture. The shift of

© 1990 by Elsevier Science Publishing Co., Inc.
Global Atmospheric Change and Public Health
James C. White, Editor

the temperate zone to higher latitudes will mean that central United States and southern Canada, which today serve as the "bread basket" for the world production of grains, may be too warm for the growth of winter grains. Whether the fertility of more northern lands will be able to replace the productivity of the Midwest is not certain. Other major food crops, particularly rice, will be intolerant of the expected warming. Rice today is a major food staple of the world.

With the global increase of temperature there will be an increase in rainfall, but there are expectations of large changes in distribution of the precipitation. Coastal areas are expected to receive the major share of the increased rain while central continental areas may suffer drought. Weather modeling can only be done for large areas of land, so predictions of what may happen in specific regions is still highly uncertain. It is expected that the central United States not only will be hotter but also much drier. The deserts of our southwest will expand to encompass the central states, the grain belt. The monsoons in Asia are expected to fail and, with that, much of the agriculture of that heavily populated region may suffer. In sub-Saharan Africa changing patterns of precipitation are likely to have greater effects upon crop production than the rise in temperature. There has been an overall tendency for lower rainfall and a later start of the rains, which is expected to continue (6); the latter effect will exacerbate the drought.

In addition to the changes in distribution of rainfall, the increased global temperature will melt snows and glaciers on mountains in temperate zones which now serve as reservoirs of water gradually released throughout the summer to feed streams and water tables. The warmer climates will either prevent accumulations of snow and ice at high altitudes or melt these prematurely, resulting in loss of water supplies during summers and increased erosion from the accelerated runoff.

Destruction of forests, which is taking place at an unprecedented rate today, leaves large areas bare of biomass which can protect the land from erosion by rain and wind. Deforestation is leading to more crop-damaging floods. Each year the world's farmers must feed 88 million more people with an estimated 24 billion fewer tons of topsoil (1). The loss of topsoil together with temperature increases and changes in rainfall will result in large areas of desertification of fertile agricultural lands.

The world's arable lands are no longer increasing to meet the needs of the growing population. Croplands are rapidly disappearing for nonfarm uses. Much disappears under concrete as roads criss-cross the landscape and urban developments pre-empt fertile lands. Although not an effect of climate change, the reduction in croplands will be an increasingly important factor affecting food supplies.

Global warming will cause a rise of sea levels from their thermal expansion and the melting of glacier and polar ice caps. The expected rise in the sea level is estimated at 1.0 meter in the next 50 to 100 years (7,8). Though this may seem an innocuous rise, it will result in flooding of many low-lying coastal areas. Coastal regions of continents, as well as low islands, will be inundated. Rich delta lands, such as portions of Louisiana and the fertile Nile delta, will disappear. In Bangladesh, about one-third of the land mass will be submerged, displacing millions of people. In Egypt, where only 4% of the land can be cultivated, food production could drop and 8.5 million people could be forced from their homes (9). In these already crowded countries, there is no place for the displaced people to go and no alternative land on which to grow crops.

The rising sea will also cause salinization of water tables beyond the submerged areas, adding this destructive factor to agriculture. Large areas of wetlands that nourish the world's fisheries would also be destroyed. The combination of displaced populations and reduced arable lands would compound the problem of feeding an increasing world population.

Atmospheric pollution is expected to directly damage agriculture. Air pollutants, such as oxides of nitrogen and sulfur which are the major constituents in acid rain, as well as ammonia, hydrogen sulfide, and dimethyl sulfide are toxic to plants, killing forests and poisoning fresh water in lakes and streams. The acidification of the environment threatens large areas of Europe and North America. Central Europe is currently receiving more than one gram of sulfur on every square meter of ground each year (10). A seven-year U.S. government study reports that air pollution is reducing U.S. crop output by at least five percent (1). Western Europe, with similar dependence on automobiles and fossil fuels, will suffer similar crop losses, while eastern Europe and China, with their heavy dependence on coal, may have even larger crop losses from air pollutants (1).

Man-made sources of nitrogen oxides, in combination with both natural and industrial hydrocarbons, are thought to be responsible for the creation of toxic levels of tropospheric ozone which contribute to smog. Levels of ozone are observed over eastern United States during summer at concentrations high enough to cause damage to crops and vegetation (11,12).

In addition to direct harmful effects on humans, UV-B radiations are also directly toxic to plants. Besides adversely affecting agriculture, UV-B can penetrate several meters into ocean waters. The cold polar surface waters below the ozone holes are especially rich in phytoplankton which are the start of the aquatic food chain. Phytoplankton are especially sensitive to UV-B radiations, so there is the potential for a severe reduction in marine food sources.

Are there likely to be compensating effects of global climate and environmental changes that will cancel the litany of disasterous effects that I have been listing? Since photosynthesis should be favored by an increase in carbon dioxide in the atmosphere, will this stimulate plant growth? The answers are not yet in. Some plants may grow faster but the effects on food crops are still uncertain.

With increased dryness and temperature, growth inhibition from high carbon dioxide levels may occur. The increase in global temperature should increase the potential for agriculture at higher latitudes, but whether the other effects we have discussed may limit food production despite more favorable growing temperatures is not yet known.

In recent times technology has come to our rescue, but it seems dubious that it can extricate us from the problems I have been enumerating. Paleontology indicates that plants and animals are highly adaptable to environmental changes provided the changes occur slowly enough. But genetic changes generally require thousands of years for the adaptations needed for the environmental changes which may occur in the next few decades. It seems possible that some grains and plants may be developed through bioengineering that are better adapted to hotter or cooler climates and to greater salinity of soils, but widespread improvements in irrigation and in chemical fertilizers that can cope with the arid deserts and the loss of topsoils seem unlikely. The loss of croplands and the movements and increases in populations seem inevitable.

Though many of the changes are already entrained and irreversible, prompt action to slow population growth, reverse environmental pollution and deterioration of the planet, and increase investment in agriculture could still prevent the worst disaster confronting the world, second only to a potential global nuclear war.

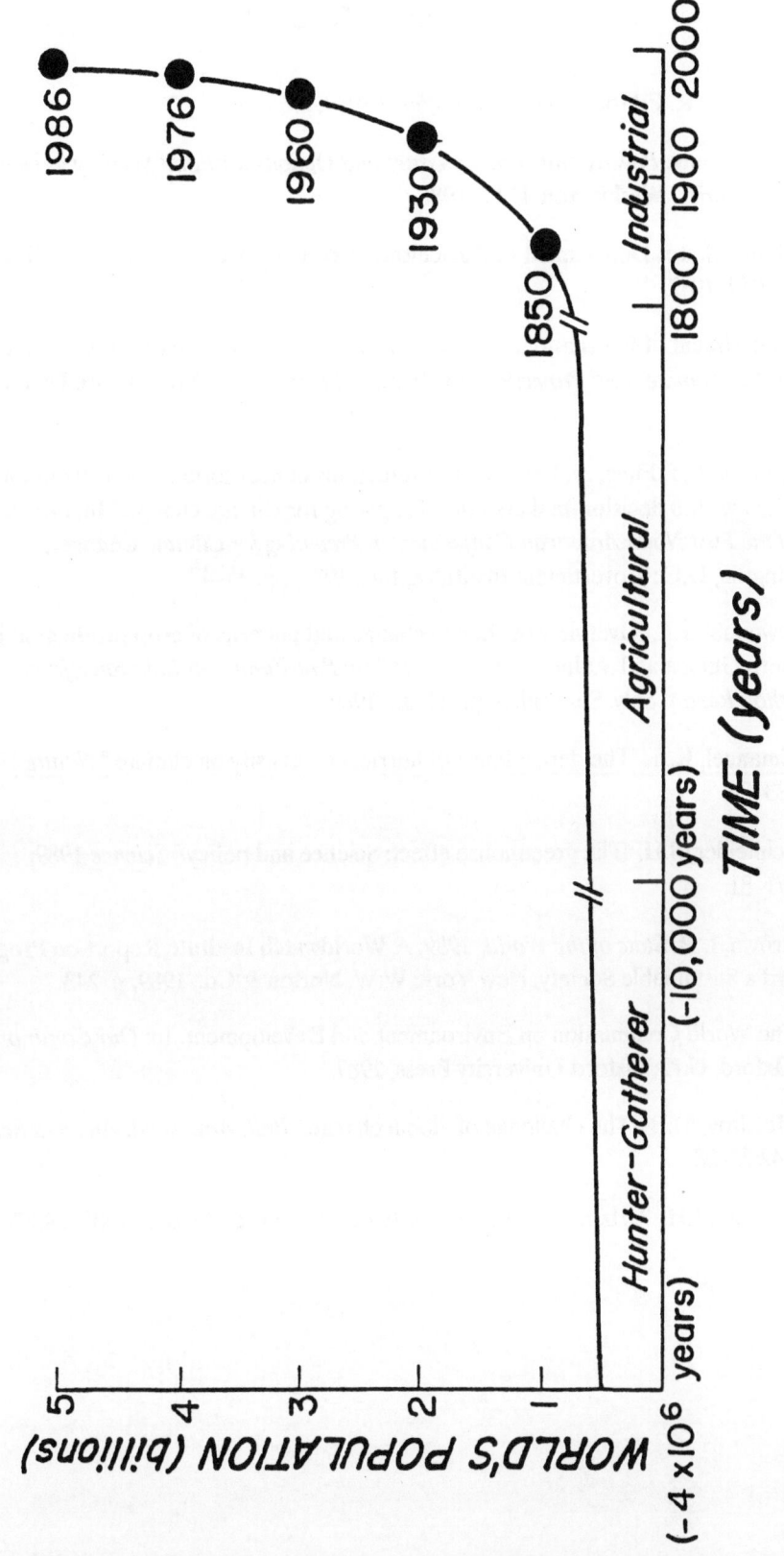

Fig. 1

REFERENCES

1.   Brown, L.R. Editor's Page. *World Watch* 1989; 2:2.

2.   World Bank. *Poverty and Hunger: Issues and Options for Food Security in Developing Countries*. Washington, D.C.: 1986.

3.   United States Department of Agriculture, Economic Research Service. *World Grain 1950–1987*.

4.   Department of International Economic and Social Affairs. *World Population Prospects: Estimates and Projections as Assessed in 1984*. New York: United Nations, 1986.

5.   Hansen, J., I. Fung, A. Lacis et al. "Predictions of near term climate evolution: What can we tell decision-makers now? Preparing for climate change." In: *Proceedings of the First North American Conference on Preparing for Climate Changes*. Washington, D.C.: Government Institutes, Inc., 1987, pp. 35–47.

6.   Owonubi, J.J., "Overview of climate change and patterns of crop production in the Northern Guinea and Sudan Savanna zones." In: *Pan-Earth Sub-Saharan Africa Workshop Report*. Saly, Senegal, Sept. 11–15, 1989.

7.   Emanuel, K.A. "The dependence of hurricane intensity on climate." *Nature* 1987; 326:483–85.

8.   Schneider, S.H. "The greenhouse effect: Science and policy." *Science* 1989; 243:771–81.

9.   Brown, L.R. *State of the World, 1989*. A Worldwatch Institute Report on Progress Toward a Sustainable Society. New York: W.W. Norton & Co., 1989, p. 243.

10. The World Commission on Environment and Development. In: *Our Common Future*. Oxford, U.K.: Oxford University Press, 1987.

11. McElroy, M.B. "The challenge of global change." *Bull. Am. Acad. Arts and Sciences* 1989; 42:25–38.

12. Seinfeld, J.H. "Urban air pollution: State of the science." *Science* 1989; 243:745–752.

CATARACTS AND ULTRAVIOLET LIGHT

Hugh R. Taylor

Dana Center for Preventive Ophthalmology
The Wilmer Institute
Johns Hopkins University
Baltimore, MD 21205

Introduction

A broad spectrum of electromagnetic radiation comes from the sun. The eye is a unique organ that is specially designed to respond exquisitely to some wavelengths of radiation, namely, visible radiation (440-700nm). The cornea and the lens transmit most of the visible radiation through to the retina where we perceive it as formed images. The cornea, the front part of the eye, absorbs almost all the radiation below 290nm, protecting the rest of the eye from the shorter wavelength ultraviolet radiation (UV-C).

The lens filters almost all the radiation between 290nm and 400nm (UV-B and UV-A) and so very little of this radiation is transmitted to the retina at the back of the eye. Visible light, above 400nm, is freely transmitted to the back of the eye. Radiation can only damage a tissue if it is absorbed by the tissue, so that one could anticipate that there would be an association between UV radiation and damage to the eye, particularly the lens and cornea.

The action spectrum of the eye shows there is a very high sensitivity of the ocular tissues to radiation below 290nm. The eye is also very sensitive to radiation between 290nm and 320nm, but beyond this the sensitivity falls progressively, although sufficient visible light can burn the retina. It has been known that people who watch a solar eclipse will develop eclipse blindness. A combination of the sudden flash of bright light and the prolonged watching of the eclipse result in the image of the sun or the eclipse being burned into the retina. This is due to the visible light and is like burning a hole in paper with a magnifying lens. Intense exposure to xenon arcs or lasers can also cause retinal burns.

Photokeratitis

It is well known that one can get snow blindness, or acute photokeratitis. This is seen in bright conditions over snowy surfaces such as in the Antarctic or Arctic areas or even

Published 1990 by Elsevier Science Publishing Co., Inc.
Global Atmospheric Change and Public Health
James C. White, Editor

in alpine areas, particularly during spring skiing. Photokeratitis is sunburn of the eye and affects the cornea. Typically, after a day of skiing with prolonged exposure, one may develop red, irritable and sore eyes. With careful examination, one can see staining of the surface of the eye that indicates the damage to the cornea. This is mainly due to exposure to UV-B, but it can also occur with UV-A and UV-C. It commonly occurs following flash burns in people using arc welders.

Photokeratitis clearly occurs with current UV-B levels, especially over bright snowy surfaces. There is also evidence that we are very close to the threshold level for photokeratitis during summer months, especially at the beach. We still do not know what level of depletion of stratospheric ozone and accompanying increase in ambient UV-B can be tolerated before photokeratitis becomes a much more widespread and common problem, but it is a serious possibility and something that is worthy of further investigation.

Cataract

Cataract causes half of all the blindness in the world today. There are about 20 million people who are blinded by cataract and more than 1.25 million cataract operations done each year in the United States alone. A cataract occurs when the normally clear lens in the eye becomes cloudy and scatters light, thereby reducing vision.

There are a number of mechanisms by which UV-B might cause cataract. The lens is made of proteins with many amino acid residues, such as tryptophane, that are particularly susceptible to photo-oxidation. UV-B can generate activated oxygen radicals which can oxidize lens proteins. UV-B radiation can also affect membrane transport (Na/K ATPase). Finally, UV-B may damage the nucleic acids in the lens epithelium.

There have been a number of studies that have suggested a linkage between UV-B, or sunlight exposure, and cataract. Some data came from the National Health and Nutrition Survey in the United States, which indicated that those who have lived in sunny states for at least half their lives had a higher risk of cataracts than people who lived in states that were not so sunny (1). Other studies done in Australian aborigines showed that with increasing levels of ambient UV-B radiation there was a progressive increase in the amount of cataract (2).

These and other similar studies were all ecologic in nature. They all assumed that if you lived in a particular place you received that much radiation. As it turns out for ocular exposure to UV-B, this is only a very rough approximation and can be quite misleading. They also used indirect measures of UV-B exposure, such as latitude or hours of sunlight in a given area. People doing the same work at the same place at the same time may have very different ocular exposures depending on hat and eyeglass use and other factors (3). If these factors are not accounted for, individual ocular exposure cannot be determined with any accuracy.

We were interested in accurately assessing ocular UV-B exposure under a variety of conditions. Initially we used miniaturized UV-B meters placed in manikin head forms. We also used polysulfone film that responds in a linear way to UV-B exposure. We were able to determine the proportion of ambient UV-B light that actually reached the eye under a variety of conditions, including over water and over grass (3). Next, we studied a group of watermen from the Eastern Shore of Maryland who worked at a variety of different water-based activities, such as crabbing, oystering, clamming and fishing. We monitored their UV-B exposure at different times of the year and while they were doing different types of work to establish what we termed the ocular ambient exposure ratio — the fraction of UV-B that reaches the cornea.

The impact of spectacles and sunglasses on ocular exposure was also assessed in addition to measuring the UV transmission characteristics of spectacle lenses themselves (4). Typical plastic lenses reduce ocular exposure by more than 90%. Glass lenses were not as effective and gave only about 80% protection.

The use of protective eye wear and hats has a major impact on exposure. People who were predominantly indoor workers have a typical exposure of 4 units (Maryland Sun Years 10-4). The typical waterman who worked outside without any ocular protection had an ocular exposure 18 times higher (72 units). If he wore a hat, his ocular exposure was cut almost in half to 47 units. If he just wore glasses, it was cut to 17 units. However, if he wore both a hat and glasses, his ocular exposure was reduced to only 8 units. Even though he is outside all day, his exposure was now only twice that of someone who was spending essentially all day inside. So there is a very dramatic impact in terms of ocular protection.

After obtaining a history of job and leisure time exposure, individual personal ocular exposures could be calculated using published data on ambient UV-B levels, the ocular ambient exposure ratio and correcting for spectacle use and so forth.

Maryland Waterman Study

We undertook an epidemiological study of the watermen who worked on Chesapeake Bay (5). They were selected for several reasons: they work year-round; they are in the open exposed to the 360° horizon; and they have stable occupation practices so one could determine lifetime exposure. We found their total exposure or skin exposure remained fairly constant up until about age 65, after which time they started to reduce the amount of time spent working outside.

Ocular exposure, however, showed a more steady reduction over time, possibly related to the use of bifocal glasses as they developed presbyopia.

There are three major types of cataract: nuclear cataract that occurs in the center of the lens; cortical cataract that occurs in the outer layers of the lens; and posterior subcapsular cataract (PSC) which occurs in the skin on the back of the lens. These different sorts of cataracts appear to have different etiologies, although all three types of cataracts show a progressive increase in prevalence and severity with increasing age.

Logistic regression analysis controlling for age showed a strong correlation between cumulative UV-B exposure and cortical cataract (6). This analysis indicates that a doubling of UV-B exposure will increase the risk of developing cortical cataracts by 60%. Conversely, if the ocular exposure to UV-B is halved, the risk is reduced by 40%. A second analysis examined the effects of annual average ocular exposure to UV-B. Those in the top quartile of annual ocular exposure had more than three times the risk of developing cortical cataract compared to people in the bottom quartile. These associations were only seen with cortical cataracts and were not seen with nuclear cataracts.

A third analysis compared the average exposure at each year of life for people who did not have cortical cataracts to those people who did. This showed that, in every year of life, people with cortical cataracts had a significantly higher ocular exposure to UV-B as compared to people who did not have cortical cataracts. This difference averaged about 20% for each year of life. There was no indication of any safe exposure or threshold of UV-B exposure nor any safe period for exposure.

The population-based waterman study found relatively few people with posterior subcapsular cataracts (PSC). To examine the factors associated with PSC, a separate case-control was done examining patients who had undergone surgery for PSC and comparing them to age-matched controls (7). Again, cumulative UV-B exposure was significantly associated with the presence of PSC. These two studies showed that two of the three major types of cataract are clearly linked with UV-B exposure.

Corneal Disease

Pterygium, a fleshy growth that grows out onto the cornea, is a common eye disease in sunny areas. At times it can affect vision, but it also can be very irritating and is cosmetically unsightly. In the waterman study, those in the upper quartile of UV exposure had more than three times the risk of developing pterygium compared to those in the lower quartile (8). Another more obscure corneal condition called climatic droplet keratopathy was also strongly related to UV exposure.

Macular Degeneration

Age-related macular degeneration is the leading cause of blindness in people over the age of 65 in the U.S. and results in an irreversible vision loss. In macular degeneration, the area of the retina responsible for fine, central vision becomes scarred although peripheral vision is retained.

It has been suggested that macular degeneration may be related to exposure to sunlight in some way and we examined its association with UV (9). However, we found there was no association with levels of exposure to either UV-B or UV-A. The risk of developing macular degeneration for those in the top quartile of ocular exposure was not significantly different compared to those in the bottom quartile. Other analyses of the waterman data also failed to show an association between UV-B and macular degeneration. There are some suggestions that macular degeneration may be associated with exposure to blue, or visible, light and that is worthy of further study.

Summary

These various studies have shown that UV-B exposure is clearly associated with cortical and posterior subcapsular cataract. This association is linear in that there is a progressive increase in the risk of developing cataract with increasing exposure. The data clearly indicate that a little bit of UV-B exposure is bad for you and a lot of exposure is worse.

The implication of these findings is that, if you are out in the sunshine and there is enough ambient UV-B radiation to cause sunburn, you should protect your eyes. A hat will reduce the ocular exposure by half; sunglasses will reduce the exposure 80-100% depending on the type of lens and the shape. The effect of the hat and glasses are additive; close-fitting glasses with UV absorbing lenses and a hat will give you the best protection.

References

1. Hiller, R., Giacometti, L. and Yuen, K., 1977. "Sunlight and cataract: an epidemiologic investigation." *American Journal of Epidemiology* 105:450–459.

2. Hollows, F. and Moran, D., 1981. "Cataract – the ultraviolet risk factor." Lancet 2:1249–1250.

3. Rosenthal, F.S., Phoon, C., Bakalian, A.E. and Taylor, H.R., 1988. "The ocular dose of ultraviolet radiation in outdoor workers." *Investigative Ophthalmology and Visual Science* 29:64.

4. Rosenthal, F.S., Bakalian, A.E., Changqi, L. and Taylor, H.R., 1988. "The effect of sunglasses on ocular exposure to ultraviolet radiation." *American Journal of Public Health* 78:72–74.

5. Taylor, H.R., in press. "Ultraviolet radiation and the eye." *Transactions of the American Ophthalmological Society*.

6. Taylor, H.R., West, S.K., Rosenthal, F.S., Munoz, B., Newland, H.S., Abbey, H. and Emmett, E.A., 1988. "Effect of ultraviolet radiation on cataract formation." *New England Journal of Medicine* 319:1429–1433.

7. Bochow, T.W., West, S.K., Azar, A., Munoz, B., Sommer, A. and Taylor, H.R., 1989. "Ultraviolet light exposure and risk of posterior subcapsular cataracts." *Archives of Ophthalmology* 107:369–372.

8. Taylor, H.R., West, S.K., Rosenthal, F.S., Munoz, B., Newland, H.S. and Emmett, E.A., 1989. "Corneal changes associated with chronic ultraviolet radiation." *Archives of Ophthalmology* 107:1481–1484.

9. West, S.K., Rosenthal, F.S., Bressler, N.M., Bressler, S.B., Munoz, B., Fine, S.L. and Taylor, H.R., 1989. "Exposure to sunlight and other risk factors for age-related macular degeneration." *Archives of Ophthalmology* 107:875–879.

SKIN CANCER AND ULTRAVIOLET LIGHT: RISK ESTIMATES DUE TO OZONE DEPLETION

Janice Longstreth

Division of Science Policy
Clement Associates
9300 Lee Highway
Fairfax, VA 22031-1207

Introduction

Stratospheric ozone depletion poses a threat to human health principally via the increase in ambient ultraviolet B (UV-B) radiation which it permits, but also potentially through a contribution to air pollution in the troposphere. UV-B has many direct impacts on the human body; most of them are undesirable but some, such as the generation of the active form of Vitamin D, are useful. It is probably a safe generalization, however, that the detrimental effects of UV-B exposures far outweigh the beneficial ones.

The direct detrimental effects of UV-B on human health include not only an increase in skin cancer but also impacts on the eye (e.g. cataracts) and on the immune system. The focus of this presentation is skin cancer. For detailed reviews of the subject, the reader is referred to a recent article and monograph (Kripke et al., 1989; USEPA, 1987).

In order to evaluate the potential impact of ozone depletion on skin cancer, one has to appreciate not only how ozone depletion can modify the amount of UV-B reaching the earth's surface but also the current dose-response relationship between UV-B exposure and skin cancer. This, in turn, requires an understanding of how the flux of UV-B currently varies by location and time.

The earth's atmosphere has two ozone layers, one very close to the surface, termed the tropospheric ozone layer, and the second extending from approximately 15 to 50 km above the surface, termed the stratospheric ozone layer. The stratospheric ozone layer contains about ten times the concentration of ozone as that in the troposphere; the two layers combined act as a protective shield, preventing a significant portion of the sun's ultraviolet radiation (UVR) from hitting the earth. This protection is somewhat selective, for the shorter wavelengths of UVR are absorbed preferentially. The shortest wavelength region of UVR is 200 to 290 nm and is termed UV-C; virtually all of it is ab-

Published 1990 by Elsevier Science Publishing Co., Inc.
Global Atmospheric Change and Public Health
James C. White, Editor

sorbed by the ozone layer. UV-B, 290 to 320 nm, is only partially absorbed, and UV-A, 320 to 400 nm, is not absorbed at all.

The UV-B incident on the earth's surface varies tremendously in quantity and quality depending on time and location. This is partly due to the differential absorption by wavelength of UVR discussed above, but also because the pathlength that solar radiation takes through the atmosphere varies temporally and geographically. For instance, energy entering at the equator has a shorter pathlength to the surface than energy incident at the poles. Figure 1 shows an example of how the amount of UVR varies by wavelength depending on the month.[1] Note that the variability is greater with the shorter wavelengths.

With depletion of the ozone layer, more UV-B will reach the earth whereas the amount of UV-A will remain unchanged and UV-C is still expected to be completely absorbed. Of the UV-B and UV-A that reach the earth, UV-B has the most biologic activity per unit of energy. Figure 2 shows the absorption spectra (between 260 and 340 nm) of a number of important macromolecules which reside in the skin. The light portion of the figure indicates those portions of UVR that are present in natural sunlight and which are therefore relevant to this discussion. Note that, with the possible exception of melanin, the absorptive capacity of these molecules is much greater in the UV-B region than in the UV-A region (i.e. above 320 nm), but that DNA and tryptophan absorb almost equally well in the relevant wavelengths.

The exact mechanism by which UV-B induces skin cancer in humans is unknown. Data from experimental systems using animals suggest that the induction of pyrimidine dimers in the DNA may be important (Ley et al., 1989). In addition, a human genetic defect, xeroderma pigmentosum, results in an inability to repair UV-B induced damage to DNA; it is also associated with a significantly increased risk of skin cancer (Kraemer et al., 1984).

Skin Cancer and Sunlight

Skin cancer is the most common neoplasm in white populations and is believed to be increasing dramatically. Unfortunately, because it is so prevalent, we do not have good historical data for the most common tumors.

---

1   Data are estimates generated by the NASA UV Satellite model designed by
     Serafino and Frederick (1986) and modified to run on a PC by Pitcher (1988).

Historically skin cancer has been divided into two forms: non-melanoma skin cancer (NMSC) which affects the principal cell type of the skin, the keratinocyte, and cutaneous melanoma (CM) which affects the pigment-producing cell, the melanocyte. NMSC is still further divided into basal cell and squamous cell carcinoma (BCC and SCC, respectively). CM is the most dangerous form of skin cancer; currently it is estimated that about 25 percent of those who develop melanoma die from it in the U.S. SCC and BCC are much less dangerous with a combined mortality of less than 1 percent, most of which is due to SCC.

Non-Melanoma Skin Cancer

Data gathering on NMSC is very poor; the U.S. has not assessed the incidence since 1977. The rate of increase in whites compared to 1970 was 15 to 20 percent. In 1977, Scotto and his colleagues at the National Cancer Institute (NCI) (1982) estimated that between 400,000 and 500,000 individuals developed NMSC annually. For SCC, this represented age-adjusted incidence rates in males and females, respectively, of 65.4 and 23.6 per 100,000. A recent report in the *JAMA* (Glass and Hoover, 1989) indicated an age-adjusted incidence rate in males for the period 1960-1986 of 106.1 per 100,000 for the same tumor. In the latter study, the study group was individuals in a health maintenance organization (HMO) in Oregon, whereas the Scotto et al. (1982) study derived its data from two special surveys performed by Scotto et al. at four and seven study areas. Interestingly, when Glass and Hoover (1989) compared their data to that of Scotto et al. (1982) for the same time period and for the same general area of the country, they found that their age-adjusted incidence was almost twice that observed by the NCI group. As they point out, there are two possible explanations for this discrepancy. In the HMO all skin lesions were submitted for pathology so that there would be virtually 100-percent ascertainment, whereas the NCI study was based on surveys of individuals diagnosed with these tumors which might have missed skin lesions not submitted for pathology. Thus the HMO data might more accurately reflect the true incidence of SCC in the population. The other explanation for the discrepancy is that it could be due to differences in the populations studied, as individuals in an HMO are probably self-selected for high socioeconomic status (SES), and high SES is a risk factor for these tumors. Nevertheless, these data are alarming for they indicate, at the very least, that there is a major segment of the population who is at far greater risk than would have been concluded on the basis of the DHHS study (Glass and Hoover, 1989). It should be noted also that the report of Glass and Hoover (1989) only presented data on squamous cell carcinoma, the relatively less prevalent tumor. If the discrepancy observed with SCC holds true for BCC, the number of individuals estimated from the Scotto et al. (1982) study to annually get non-melanoma skin cancer could also be a two-fold underestimate. It is of interest to note that, for the period of 1982 to 1986 in a population-based skin cancer survey in Victoria, Australia, the age-standardized incidence rates per 100,000 for BCC and SCC were 671 and 201, respectively (Marks et al., 1989). Thus in an area of high solar insolation, the incidence for SCC was twice what it was in the Oregon HMO.

Epidemiologic evidence clearly indicates an association between both BCC and SCC with increased cumulative exposure to sunlight. Individuals at highest risk are those who spend the greatest amount of time outdoors (e.g. farmers and fishermen), and sites receiving the most exposure (e.g. face and hands) show the greatest incidence. In addition, light-skinned individuals are at greater risk than dark-skinned individuals, and there is a latitude gradient in the U.S. so individuals in the South are at higher risk than those in the North.

Cutaneous Melanoma

Cutaneous melanoma is a far more virulent disease than BCC or SCC. The current mortality rate in white populations is about 25 percent. The data gathered on CM in the U.S. are much more complete than that for NMSC. In the U.S., during the period 1977-1985, CM increased 57 percent in whites and decreased 10 percent in blacks. The current age-adjusted incidence rates for males and females in the U.S. based on the 1982-1986 NCI SEER data are 10.8 and 8.3 per 100,000, respectively (NCI, 1988); the recent report from Glass and Hoover (1989) found age-adjusted incidence rates for males and females, respectively, for approximately the same period of 20.1 and 17.0 per 100,000 — again about twice that found by NCI.

The relationship between sunlight and CM is less clear-cut than that for NMSC. A number of reasons contribute to this lack of clarity. First, although CM results from the neoplastic transformation of melanocytes, there are a variety of forms of CM which have dissimilar relationships to sunlight or UV-B exposure. The most common classification of CM includes the following entities:

superficial spreading melanoma (SSM)
nodular melanoma (NM)
lentigo maligna melanoma (LMM)
acral lentigenous melanoma (ALM)

These various types appear to have very different relationships to sunlight exposure, and most of the population-based studies (e.g. ecologic epidemiology showing latitude gradients) cannot take these differences into account. Of the four types listed above, ALM shows no relationship to sunlight, LMM shows a reasonably strong relationship to cumulative exposure, and NM and SSM show a somewhat paradoxical relationship. Unlike NMSC and LMM, cumulative exposure is not very strongly associated with a high risk of SSM and NM. However, individuals with little pigment (i.e. fair skin, blue eyes, light colored hair) are at greatest risk. Freckling and nevus formation, which are associated with sun exposure, are also risk factors. In addition, individuals living closer to the equator are at higher risk than those whose homes are removed from the equator.

The difference between these tumor types appears to be that, in the case of NMSC and LMM, the cumulative dose is critical whereas, in the case of NM and SSM, some sort of peak dose or overwhelming of a threshold is critical.

Evidence that UV-B is the Active Component of Sunlight

Action Spectra for NMSC in Animals

Animal experiments indicate UV-B to be the most active portion of the solar spectrum in inducing NMSC; the tumors induced by UV-B in animals are fibrosarcomas and SCC. An action spectrum for NMSC induction in mice is presented in Figure 3; it is derived from work of Slaper (1987) and his colleagues presented in a recent UNEP report (van der Leun, 1989). BCC is rarely, if ever, induced in animals, so neither a dose-response relationship nor an action spectrum has been determined. However, since the site distribution of BCC and the populations at risk are so similar to those observed for SCC (USEPA, 1987), UV-B is generally accepted as the principal etiologic agent in humans for BCC as well.

UV-B Induction of Melanoma in Opossums and Fish

Recently a new animal model has been reported which should prove useful to the examination of the role of UV-B in melanoma induction. Monodelphis domestica is a small South American opossum which has been used by Ley and his colleagues to examine a number of interesting questions in photobiology (Ley, 1985; Ley and Applegate, 1985, 1989). Recently this group has reported on the induction of malignant melanomas in these animals following almost 70 weeks of irradiation with UV light (280-400 nm) (Ley et al., 1989). Animals were exposed to UVR three times a week (on shaved skin) at an average exposure of 250 J/m^2 (about one-half of the minimal erythemal dose). The tumors were first noted at 100 weeks; however, a precursor lesion, melanocytic hyperplasia was discovered much earlier (at 12-14 weeks).

One of the interesting traits of this opossum is the presence of a photoreactivation repair pathway for DNA damage. Photoreactivation repair is a light-dependent process wherein cyclobutane-type pyrimidine dimers (such as those induced by UVR) can be repaired *in situ* simply by exposing the animals to light of the appropriate wavelength (300-500 nm) following the damage-inducing irradiation with UV-B. Using this technique, Ley and his colleagues have been able to demonstrate that the melanocytic hyperplasia is reversible upon photoreactivation, leading them to conclude that DNA damage is an important step in melanoma formation.

In a second recently described model system in fish, Setlow and his colleagues (1989) have examined the role of UVR in the induction of melanomas in hybrids between the platyfish and swordtail. Using UVR from sunlamps filtered to allow passage of radiation > 290 nm or radiation > 304 nm, these investigators found that both single and multiple dose regimens involving relatively low doses of UVR (850 J/m^2 of radiation > 304 nm was the lowest dose to cause tumors) would lead to a significantly increased incidence of melanomas in these animals.

Xeroderma Pigmentosum

Although there is a relative paucity of experimental (i.e. non-epidemiologic) data from humans, one of the strongest pieces of evidence which supports the hypothesis that exposure to UV-B plays a role in melanoma etiology comes from studies of a rare human genetic defect: xeroderma pigmentosum (XP). Cells from individuals with this disease show a hypersusceptibility to cell-killing and mutagenesis by UVR which is associated with a defect in DNA repair. There are a number of XP phenotypes; however, the defect most often found is an inability to perform the initial step of pyrimidine dimer excision (Cleaver, 1983). XP individuals experience a variety of cutaneous manifestations including a high incidence of non-melanoma and melanoma skin cancer. Kraemer and his colleagues (1984) estimate that, before the age of 20, XP patients have a 2,000-fold increased risk of melanoma and a 4,800-fold increased risk of BCC and SCC.

Dose-Response Relationships

Non-Melanoma

Dose-response relationships for BCC and SCC can be derived from epidemiologic data (i.e. the latitude gradient described above) and, in the case of SCC, from animal data as well. In the case of the epidemiologic data, dose information can come either from Robertson-Berger meters set at the sites where the data have been gathered, or can be estimated using a NASA satellite model of UVR flux coupled with the appropriate action spectrum. Dose-response relationships for the two tumors differ somewhat, however, in that each increment of UVR is associated with a greater increase in SCC than BCC.

Melanoma

The differences in behavior vis-à-vis sun exposure between the various forms of melanoma make estimation of the impact of ozone depletion CM somewhat more difficult. The dose-response relationship between LMM and sunlight or UVR exposure appears to be that of a cumulative relationship, whereas it seems likely that some sort of intermittent relationship is required for NM and SSM — yet the data we have from which to derive dose-response relationships is drawn from studies which do not separate the tumor types. Furthermore, none of the studies which might provide dose-response relationships have been able to explore qualitative differences in dosimetry, e.g. somehow including weights to account for the fact that exposures early in life (particularly if severe) may be more important than those later on.

As a crude approximation, however, dose-response relationships can be developed using the epidemiologic information indicative of a latitude gradient and assuming that an estimate of chronic UVR exposure (such as annual dose of UV) is an appropriate surrogate for the correct dose parameter. Thus these site-specific dose estimates can be coupled with information on melanoma incidence and mortality at those sites.

Risk Estimates Due to Ozone Depletion

Derivation of risk estimates for SCC and BCC using either the animal or the epidemiologic data have produced similar conclusions (USEPA, 1987; van der Leun et al., 1989). These reports suggest that, for every 1-percent decrease in stratospheric ozone, there will be an increase in SCC of between 3 and 5 percent and an increase in BCC of between 2 and 3 percent. Similar analyses for melanoma (Longstreth, 1987) led to estimates that a 1-percent decrease in ozone would be associated with between a 1- and 2-percent increase in CM incidence and between 0.3- and 2-percent incidence in CM mortality.

On the basis of these dose-response relationships, the USEPA estimated that, were ozone depletion allowed to continue until 40-percent depletion (estimated to occur in 2075), the increase in biologically active UV would permit an additional 154 million cases of skin cancer and an additional 3.4 million deaths (this estimate was for individuals alive at the beginning of the assessment and born through 2075). If, however, the original provisions of the Montreal Protocol were put into place, it was estimated that these measures would reduce the additional skin cancer deaths to about 3 million and the number of deaths to 300,000 in that same population.

Unfortunately, the risk estimates performed in 1987 and 1988 were based on model estimates of the amount of ozone depletion using a model which indicated that ozone depletion was currently at approximately 0.5 percent. The discovery of the ozone hole (Farman et al., 1985) and then a subsequent report by NASA indicating a global average depletion of about 2.5 percent (NASA, 1988), both of which were not predicted by the models, indicated that there were serious deficiencies in the model and thus that the estimates were very likely gross underestimates. This further suggested that the provisions of the Montreal protocol were probably insufficient. Meetings in Helsinki in late 1988 led to the Helsinki declaration in which a complete phase-out of CFCs and other ozone-depleting chemicals such as methyl chloroform was proposed.

Conclusions

The USEPA is currently developing revised estimates of the number of skin cancer cases and deaths likely to result from ozone depletion with or without a complete phase-out. Additional factors to be considered in that analysis will include possible changes in the dose-response relationship for melanoma where exposure in the first 20 years of life is weighed more heavily than exposures later in life, changes in the base-case estimates for both melanoma and non-melanoma skin cancer, and changes in the starting point for ozone depletion. It seems highly likely that the risk estimates will increase. The interesting questions, however, will be what measures will be necessary to stabilize the ozone layer, how much depletion have our activities already committed us to, and how much ultraviolet radiation will reach the surface when such depletion occurs.

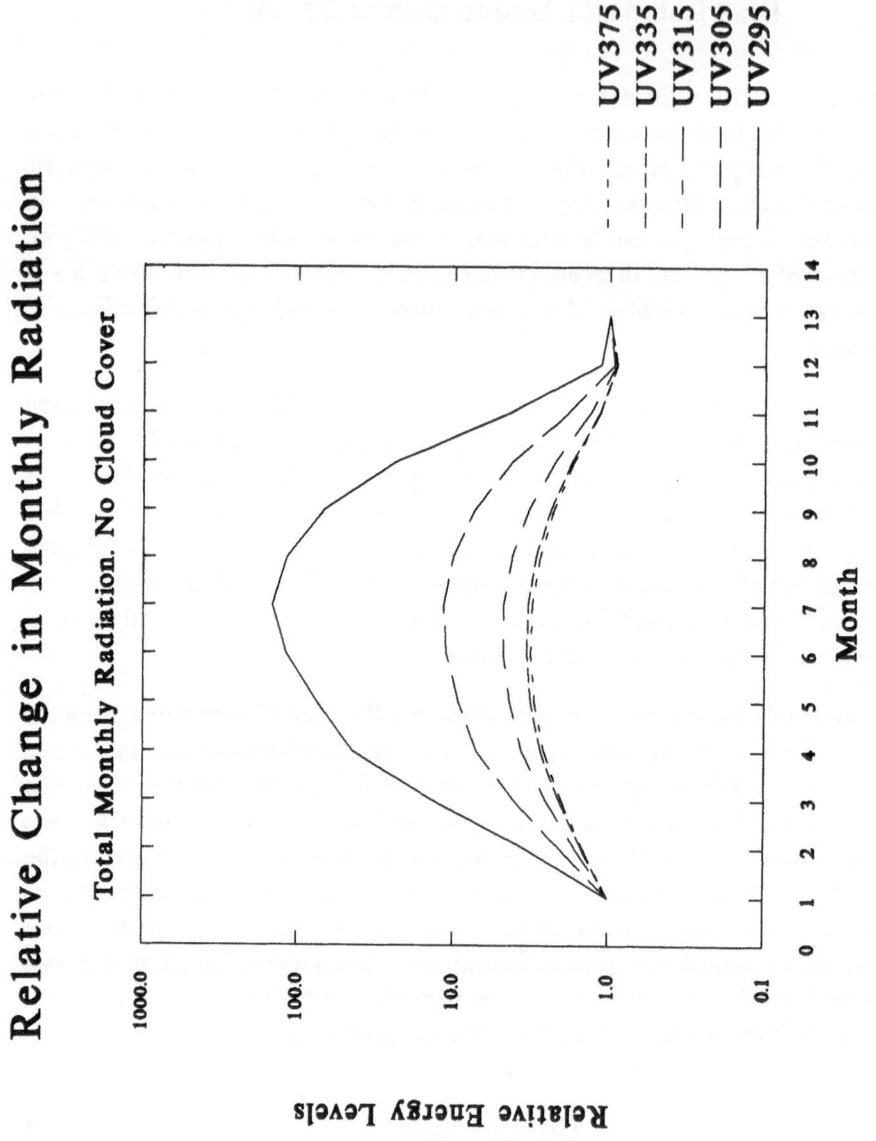
Relative Change in Monthly Radiation
Total Monthly Radiation. No Cloud Cover
Relative Energy Levels
1000.0
100.0
10.0
1.0
0.1
0 1 2 3 4 5 6 7 8 9 10 11 12 13 14
Month
UV375
UV335
UV315
UV305
UV295

Figure 1
Variation in UVR by month in Washington, D.C.

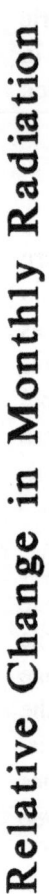

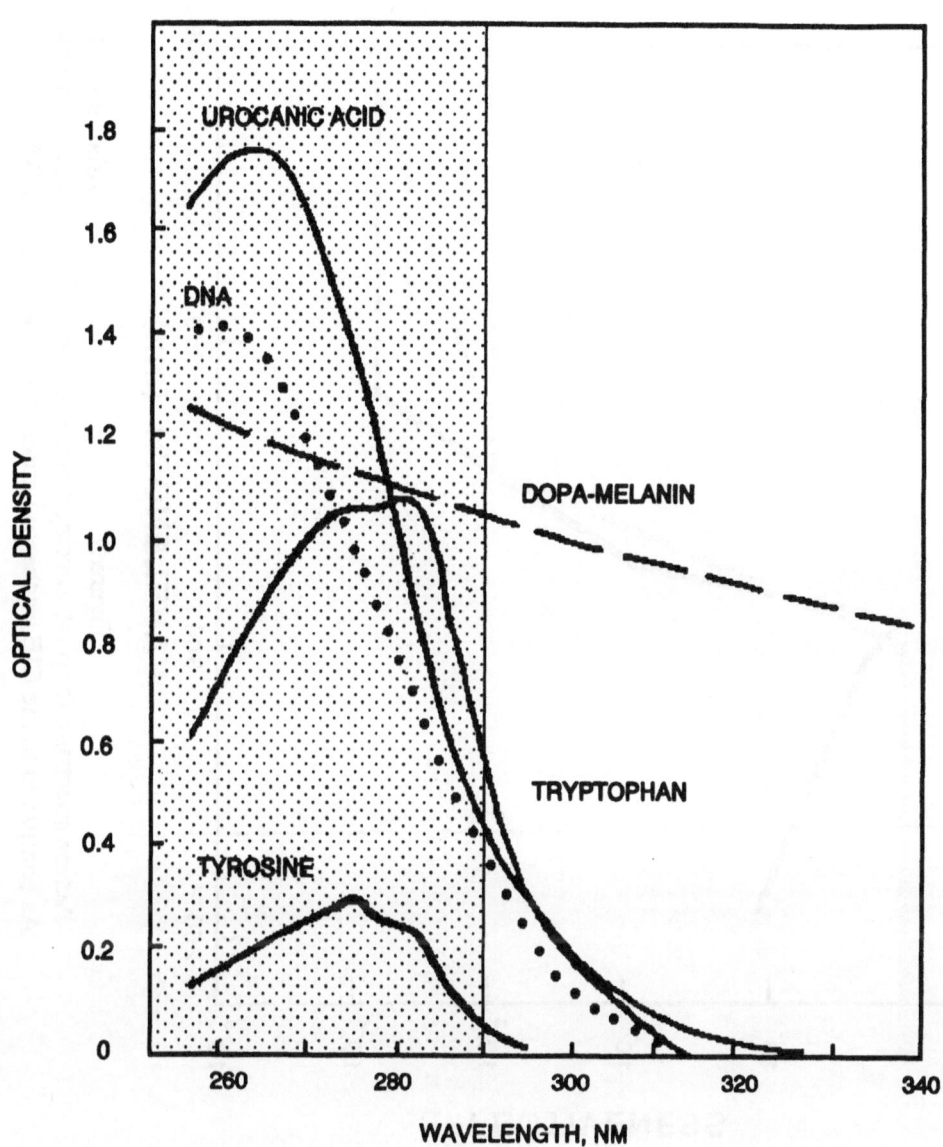

Figure 2
Absorption spectra for major epidermal chromophores

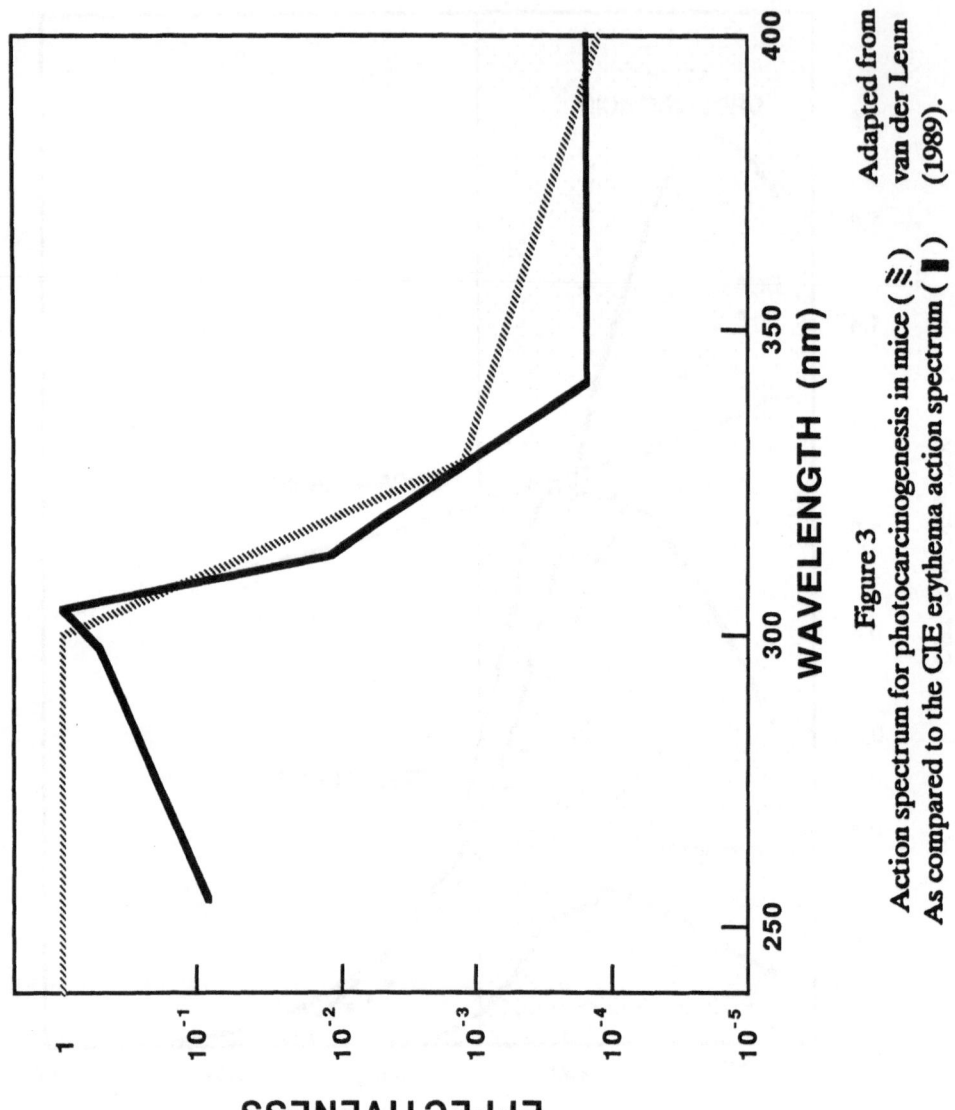

Figure 3

Action spectrum for photocarcinogenesis in mice (≋)
As compared to the CIE erythema action spectrum (▮)

Adapted from
van der Leun
(1989).

References

Cleaver, J.E. "Xeroderma pigmentosum." In *The Metabolic Basis of Inherited Disease*, Stanbury, J.B., Wyngaarden, J.B., Frederickson, D.S., Goldstein, J.S. and Brown, M.S. (eds.). 5th edition. New York: McGraw-Hill Book Co. Pp. 1227–1248, 1983.

Farman, J.C., Gardiner, H. and Shanklin, J.D. "Large losses of total ozone in Antarctica reveal seasonal ClO_x/NO_x interaction." *Nature* 315:207–211, 1985.

Glass, A.G. and Hoover, R.N. "The emerging epidemic of melanoma and squamous cell skin cancer." *JAMA* 262:2097–2100, 1989.

Kraemer, K.H., Lee, M.M. and Scotto, J. "DNA repair protects against cutaneous and internal neoplasia: evidence from xeroderma pigmentosum." *Carcinogenesis* 5:511–514, 1984.

Kripke, M.L., Pitcher, H., Longstreth, J.D. "Potential carcinogenic impacts of stratospheric ozone depletion." *Env. Carcino Rev. (J. Envir. Sci. Hlth.)* C7:53–74, 1989.

Ley, R.D. "Photorepair of pyrimidine dimers in the epidermis of the marsupial *Monodelphis domestica*." *Photochem. Photobiol.* 40:141–143, 1984.

Ley, R.D. and Applegate, L.A. "Ultraviolet radiation-induced histopathologic changes in the skin of the marsupial *Monodelphis domestica*. II. Quantitative studies of the photoreactivation of induced hyperplasia and sunburn cell formation." *J. Invest. Dermatol.* 85:365–367, 1985.

Ley, R.D. "Photoreactivation of UV-induced pyrimidine dimers and erythema in the marsupial *Monodelphis domestica*." *Proc. Natl. Acad. Sci. (U.S.A.)* 82:2409–2411, 1985.

Ley, R.D., Applegate, L.A., Padilla, R.S., Stuart, T.D. "Ultraviolet radiation-induced malignant melanoma in *Monodelphis domestica*." *Photochem. Photobiol.* 50:1–5, 1989.

Longstreth, J.D. (ed.). *Ultraviolet Radiation and Melanoma with a Special Focus on Assessing the Risks of Ozone Depletion*, EPA 400/1–87/001D, 1987.

Marks, R., Jolley, D., Dorevitch, A.P., Selwood, T.S. "The incidence of non-melanocytic skin cancers in an Australian population: results of a five-year prospective study." *Med. J. Aust.* 150:475–478, 1989.

National Aeronautics and Space Administration (NASA). *Executive Summary, Report of the Ozone Trends Panel*, March 4, 1988.

National Cancer Institute (NCI), Division of Cancer Prevention and Control. *1987 Annual Cancer Statistics Review, including Cancer Trends: 1950–1985*. National Cancer Institute, 1988.

Scotto, J., Fears, T.R., and Fraumeni, J.F., Jr. *Incidence of Nonmelanoma Skin Cancer in the United States*. National Cancer Institute, U.S. Department of Health and Human Services, Public Health Service, NIH 82–2433, 1982.

Setlow, R.B., Woodhead, A.D. and Grist, E. "Animal Model for Ultraviolet Radiation-Induced Melanoma: Platyfish-Swordtail Hybrid." *Proc. Natl. Acad. Sci.* **86**:8922–8926, 1989.

Slaper, H. *Skin Cancer and UV Exposure: Investigations on the Estimation of Risk*. Ph.D. Thesis, Utrecht, 1987.

United States Environmental Protection Agency (USEPA). *Assessing the Risks of Trace Gases that Can Modify the Stratosphere*. Washington, D.C.: Government Printing Office, EPA 400/1–87/001C, 1987.

Van der Leun, J.C., Takizawa, Y. and Longstreth, J.D. "Human Health," Chapt. 2 in *Environmental Effects Panel Report*. United Nations Environment Programme, 1989.

Acknowledgments

This work was partially supported by EPA contracts 68–02–4601, and 68–01–7289. The information presented is totally the responsibility of the author and should not be construed as Agency position or policy. The author would like to thank John Hoffman for his support, Hugh Pitcher for the development of some of the graphics, and Hugh Pitcher and C. Suzanne Lea for many helpful discussions.

RESPIRATORY EFFECTS ASSOCIATED WITH GLOBAL CLIMATE CHANGE

Lester D. Grant[1]

Environmental Criteria and Assessment Office
Office of Health and Environmental Assessment
U.S. Environmental Protection Agency
Catawba Building (MD-52)
Research Triangle Park, NC 27711

Abstract

Notable changes in global climate are projected as likely to occur during the next several decades and well into the next century due to (1) stratospheric ozone depletion caused by anthropogenic emissions of chlorofluorocarbons (CFCs), halons (bromine compounds) and other compounds, and (2) global warming due to "greenhouse" gases (such as carbon dioxide, nitrous oxide, tropospheric ozone, and CFCs). These global climate changes are expected to result in many human health and environmental impacts. Included are likely increased frequency and duration of air stagnation periods, which may contribute to more severe air pollution episodes over urban and rural areas during which more marked elevations in surface level air pollutants (ozone, sulfur dioxide, particulate matter, acid aerosols, etc.) are likely to pose increased health risks for exposed human populations. Decrements in pulmonary function, increased respiratory symptoms, impairments of lung defense mechanisms, and possibly more chronic damage to lung tissue and earlier loss of lung capacity with aging are types of respiratory effects associated with exposures to ozone; decrements in lung function, increased morbidity (e.g. higher incidence of bronchitis), and increased mortality (at sufficiently high exposure levels) are associated with exposures to sulfur dioxide, its acidic transformation products, and/or other particulate matter species. The occurrence and severity of these types of effects associated with increased air pollution levels in many areas of the world, then, will likely be exacerbated by stratospheric ozone depletion and global warming. In turn, certain feedback effects (e.g. increased tropospheric ozone) are likely to contribute to further increases in global warming. Such linkages highlight the

---

1   The views presented in this paper are those of the author and do not necessarily represent official U.S. Environmental Protection Agency positions or policies.

Published 1990 by Elsevier Science Publishing Co., Inc.
Global Atmospheric Change and Public Health
James C. White, Editor

need for national and international strategies, both to deal with emerging stratospheric ozone depletion and global warming/climate change issues and, also, with interrelated tropospheric air pollution problems.

Introduction

Both stratospheric ozone depletion and global warming (greenhouse effect) are expected to contribute to global climate change processes that are estimated as likely to have enormous future impacts on both human health and the environment. Also, both global warming and stratospheric ozone depletion are recognized as problems arising from the emissions of particular air pollutants due to man's industrial/commercial activities, the long-range transport of those pollutants and/or their atmospheric transformation products, and their typically long atmospheric residence times (ranging up to 75-100 years or more for some of the causative pollutants), as described in detail elsewhere (Villach, 1985; U.S. EPA, 1987).

More specifically, stratospheric ozone depletion is expected to result from: worldwide emissions of chlorofluorocarbons (CFCs) widely used as refrigerants and halons (bromine-containing compounds) used as flame retardants in fire extinguishers and shipboard fire suppressant systems; the migration of the CFCs and halons over the course of decades to the upper atmosphere (the stratosphere); and the catalyzing there by chlorine and bromine of destruction of stratospheric ozone. Global warming, on the other hand, is expected to occur due to trapping by so-called "greenhouse" gases of heat from incoming sunlight and reflectance of heat back from the earth's surface. Among those gases having the strongest warming effects are: carbon dioxide (CO_2) emitted largely from fossil fuel (coal, oil, etc.) combustion; methane emitted both from numerous natural sources and from various anthropogenic activities (agricultural practices, coal mining, etc.); tropospheric ozone formed by photochemical reactions between volatile organic compounds (VOCs) and nitrogen dioxide (NO_2) that are catalyzed by ultraviolet (UV-B) light; and the CFCs.

There is a growing recognition that important interrelationships exist between global climate change problems and other air pollution problems that involve not only local impacts on human health and the environment in areas proximal to the primary pollutant sources, but also impacts on much broader, regional or global scales, due to long-range transport of primary emission pollutants and/or their atmospheric transformation products (Grant, 1988). This paper discusses some illustrative examples of such interrelationships, especially as they pertain to respiratory system effects that can be projected as likely to be exacerbated by global climate change due to stratospheric ozone depletion and/or global warming. Also it should be noted that, because of the nature of the types of problems and interrelationships discussed here, there exists changing perspectives on the part of many national governments and the international community with regard to how to develop strategies by which to deal with such problems. In general, there is a shift away from mainly focusing on addressing air pollution problems on a local

scale toward the recognition of the need, also, for international cooperation in dealing
with both the sources and the ultimate, global impacts of such pollution (Grant, 1988).

Global Climate Change Impacts

Stratospheric ozone depletion is anticipated to cause increased penetration of UV-
B light to the surface of the Earth, thus causing a wide variety of potential environmental and human health impacts. The latter include increased incidence of cataracts and
other ocular effects, certain types of skin cancers, and suppression of immune system
components which might lead to increased vulnerability to some infectious diseases
and/or decreased effectiveness of vaccinations (U.S. EPA, 1987; Grant, 1988; van der
Leun, 1989). In addition, UV-B catalyzes the atmospheric formation of surface-level
(tropospheric) ozone and, as such, represents one mechanism by which stratospheric
ozone depletion may contribute to increased tropospheric ozone air pollution levels
(Gery, 1989). Also, tropospheric ozone and other surface-level air pollutants are likely
to be increased due to both stratospheric ozone depletion and global warming increasing the frequency and duration of air stagnation periods, contributing to more frequent
and longer air pollution episodes in many urban areas and higher levels of air pollutants
over rural areas of the world as well (Villach, 1985; U.S. EPA, 1987). Thus, it is not unreasonable to anticipate that global climate change will exacerbate existing air pollution
problems and consequent environmental and human health impacts. Such impacts include notable respiratory system effects associated with exposure to tropospheric ozone,
sulfur dioxide, acid aerosols and other particulate matter species.

Tropospheric Ozone Effects

Elevated levels of surface-level ozone represent one of the most serious current
United States air quality problems. There are estimated to be about 100 U.S. areas out
of compliance with our current National Ambient Air Quality Standard (NAAQS) for
ozone, which is 0.12 ppm (1 hr). This includes not only the Los Angeles area but many
other urban areas, e.g. in the eastern United States where ambient ozone levels have
generally increased during the past several years. Also many other urban areas of the
world (such as Mexico City) have serious air pollution problems involving elevated surface-level ozone and consequent human exposures.

There are several bases for concern about ozone-induced health effects, as
evaluated elsewhere (U.S. EPA, 1986a; Grant, 1988; Lippmann, 1989a). Acute exposures cause a variety of respiratory system effects. To summarize briefly, key research
findings indicate that ozone exposure for one hour at concentrations of 0.12 ppm or
above (with intermittent heavy exercise) causes transient pulmonary function effects that
result in rapid shallow breathing and decreased maximum inspiration, as well as
respiratory symptoms (e.g. cough, substernal pain, other discomfort), among sensitive
individuals tested in controlled human exposure studies. About 5-20% of the subjects

tested are found to be "responders", i.e. they have much greater than average changes in lung function or respiratory symptoms than others. No notable differences in sensitivity to ozone appear to exist among normal healthy adults, children, asthmatics, and chronic obstructive pulmonary disease (COPD) patients as far as effective ozone levels causing respiratory effects. However, the same magnitude of pulmonary decrement seen at a given ozone level may represent a medically more significant effect for asthmatics and/or COPD patients with already impaired respiratory functions than for normal, healthy individuals. Furthermore, children are of concern since they typically exercise outdoors more than adults, thereby increasing their exposure to ozone at a time when their lungs are developing. It should also be noted that exposures to ambient air mixes containing elevated ozone and other pollutants (such as acid aerosols) cause respiratory effects at somewhat lower ambient ozone levels than those seen with controlled exposures to ozone alone. In addition, the magnitude and severity of observed ozone effects increase with more prolonged exposures (e.g. for 6-7 hrs), and significant changes in lung function measurements can be observed at lower (e.g. 0.08 ppm) levels than those found to cause effects with shorter exposure durations. Ozone also increases airway hyperresponsiveness to bronchoconstrictor agents such as histamine or other chemical mediators released in response to biological agents causing allergic reactions. The types of effects noted here may contribute to decreased work performance, disruption of recreational/social activities of exposed individuals, and possibly cause some individuals (e.g. asthmatics) to use medication or to seek medical attention.

Besides the above pulmonary function decrements, experimental studies provide evidence for: (1) lung inflammation responses in humans and animals with ozone exposures; (2) indications in animals of persisting lung toxicity (e.g. altered structure, increased collagen production possibly leading to fibrosis, etc.) and more rapid decrease in lung functions with age as the result of chronic ozone exposure; and (3) impacts of ozone in animals on lung defense mechanisms (e.g. alveolar macrophages), which are consistent with observed increased susceptibility to respiratory infections. These latter types of effects are of growing concern in terms of implicating chronic ozone exposure as likely contributing to longer-term, persisting respiratory disease processes in humans; but more research is necessary to better quantify concentrations and durations of ambient ozone exposures associated with induction of such effects in humans.

The numbers of individuals in the population experiencing the above types of effects would be expected to be increased with more marked elevations and prolonged durations of exposure to surface-level ozone, projected as likely to occur because of global climate change induced by stratospheric ozone depletion and global warming. Under conditions of elevated temperatures arising from global warming processes, the respiratory effects of ozone may be further exacerbated by heat stress associated with the higher temperatures as well, but more research is needed to better characterize what might be expected. Also there exists the possibility that higher temperatures, especially when coupled with elevated humidity, may lead to increased levels of pollens and other biologic agents causing allergic respiratory system reactions in humans whose reactions are potentially exacerbated in severity by ozone-induced airway hyperresponsiveness. Temperature changes may also affect the distribution of infectious diseases. Since ozone and certain other tropospheric air pollutants increase susceptibility to respiratory infections, such changes could occur with global warming and need to be characterized further.

Sulfur Dioxide, Particulate Matter, Acid Aerosol Effects

Health effects of certain other often co-occurring pollutants may also be exacerbated due to increased frequency of air pollution episodes resulting from global climate change. More specifically, sulfur dioxide (SO_2) and particulate matter (PM) emissions from combustion of fossil fuels (coal, oil, etc.) and acidic transformation products of SO_2 have historically contributed to notable health effects in many cities and continue to do so in many areas of the world, as discussed in detail elsewhere (U.S. EPA, 1982, 1986b, 1989; Grant, 1988). Such health effects include acute pulmonary function decrements (e.g. increased airway resistance leading to decreased air flow to the lungs) seen especially in exercising asthmatics with brief (< 1 hr) SO_2 exposures as low as approximately 0.25 to 0.5 ppm. They also include more serious morbidity (e.g. increased incidence of bronchitis, other respiratory disease symptoms) and/or mortality effects associated especially with acute marked combined elevations of SO_2 and airborne particles, as occurred in famous multi-day air pollution episodes in London during the 1950s and early 1960s, pollutant elevations still currently seen in various areas of the world (e.g. in certain parts of eastern Europe). Certain pulmonary function decrements continue to be associated with acute and chronic exposures to lower ranges of SO_2 and PM ambient concentrations more typically found now in the United States and western European countries. Newer evidence (Lippmann, 1989b; U.S. EPA, 1989) tends to suggest that the morbidity and/or mortality effects may be most closely associated with sulfuric acid or other acidic sulfate transformation products of SO_2.

Again, to the extent that air pollution episodes are projected as likely to increase in response to global climate change induced by stratospheric ozone depletion and global warming, then it is reasonable to predict likely increases in numbers of individuals exposed more frequently to elevated levels of SO_2, acid aerosols, and other PM species, with the above types of consequent health impacts.

Linkages between Causative Factors Contributing to Tropospheric Air Pollution and Global Climate Change

In addition to understanding ways (such as those described above) that global climate change may exacerbate existing surface-level air pollution problems and resulting respiratory system or other health impacts, it is extremely important to recognize crucial linkages between causative factors contributing to both tropospheric air pollution and global climate change. For example, as noted above, CFCs cause stratospheric ozone depletion and contribute to global warming as well. Increased penetration of UV-B light due to stratospheric ozone depletion, in turn, is expected to increase tropospheric ozone levels with consequent impacts not only on the environment and human health, but also in terms of the tropospheric ozone (as a greenhouse gas) contributing to further global warming. These linkages can be expected to be further worsened by global warming, in turn contributing to increased tropospheric ozone levels completing yet another link in a serious, deleterious atmospheric feedback system.

Another set of salient linkages relate to fossil fuel combustion as the source of atmospheric pollutants that both cause tropospheric air pollution problems and contribute to global warming as well. That is, the combustion of fossil fuels such as coal and oil releases large amounts of CO_2 as one of the major greenhouse gases. Combustion of such fuels, without effective control of emissions from stationary or mobile sources, also results in the emissions of sulfur dioxide and particulate matter contributing to major surface-level air pollution problems as noted above. Also increased are emissions of nitrogen oxides, key precursors for the formation of surface-level ozone. Increased levels of tropospheric ozone (another greenhouse gas), in turn, are both increased by global warming and contribute to global warming, adding to the deleterious linkages between fossil fuel combustion, tropospheric air pollution and global climate change.

Conclusions

Given the interrelationships noted above, there appears to exist a sound basis by which to project that both global warming and stratospheric ozone depletion are likely to have global climate change impacts that will worsen existing surface-level air pollution problems (e.g. increased tropospheric ozone and sulfur dioxide-related pollution) and consequent human health and environmental effects of such pollution. Increased levels of tropospheric ozone and other directly emitted greenhouse gases (e.g. CO_2 and methane), in turn, are expected to contribute to further global warming. Such linkages highlight the need for developing effective national and international strategies for dealing with both emerging stratospheric ozone depletion and global warming/climate change issues and, also, interrelated tropospheric air pollution problems.

References

Gery, M.W., 1989. *Tropospheric Air Quality*, Environmental Effects Panel Report, Pursuant to Article 6 of the Montreal Protocol in Substances that Deplete the Ozone Layer. United Nations Environment Programme (UNEP), Nairobi, Kenya.

Grant, L.D., 1988. *Health Effects Issues Associated with Regional and Global Air Pollution Problems*, Conference Proceedings, The Changing Atmosphere Implications for Global Security. Report No. WHO/OMM No. 710, 1988.

Grant, L.D., R.W. Elias, R.A. Goyer, H. Olem, W.J. Nicholson, P.M. Bertsch, J.M. Davis, A.R. Flegal, K.R. Mahaffey, 1990. *Indirect Health Effects of Acidic Deposition*. Washington, DC: National Acid Precipitation Assessment Program; State-of-the-Science/Technology Paper No. 23. (In press.)

Lippmann, M., 1989a. "Health effects of ozone: a critical review." *J. Air Pollution Control Assoc.* **39**: 671-695.

Lippmann, M., 1989b. "Background in health effects of acid aerosols." In: *Symposium on Health Effects of Acid Aerosols* (October 1987). Research Triangle Park, NC. Environmental Health Perspectives **79**: 3-6.

U.S. Environmental Protection Agency, 1986a. *Air Quality Criteria for Ozone and Other Photochemical Oxidants*. Vol. I-V. Office of Health and Environmental Assessment, Environmental Criteria and Assessment Office, Research Triangle Park, NC. EPA Report No. EPA/600/8-84/020aF-eF.

U.S. Environmental Protection Agency, 1982. *Air Quality Criteria for Particulate Matter and Sulfur Oxides*. 3 v. Office of Health and Environmental Assessment, Environmental Criteria and Assessment Office, Research Triangle Park, NC. EPA Report No. EPA-600/8-82-029aF-cF.

U.S. Environmental Protection Agency, 1986b. *Second Addendum to Air Quality Criteria for Particulate Matter and Sulfur Oxides (1982): Assessment of Newly Available Health Effects Information*. Office of Health and Environmental Assessment, Environmental Criteria and Assessment Office, Research Triangle Park, NC. EPA Report No. EPA-600/8-86-020F.

U.S. Environmental Protection Agency, 1987. *Assessing the Risks of Trace Gases That Can Modify the Stratosphere*. Vol. I-V. Office of Air and Radiation, Washington, DC. EPA Report No. EPA/400/1-87/001A-E.

U.S. Environmental Protection Agency, 1989. *An Acid Aerosols Issue Paper: Health Effects and Aerometrics*. Office of Health and Environmental Assessment, Environmental Criteria and Assessment Office, Research Triangle Park, NC. EPA Report No. EPA/600/8-88-005F.

Van der Leun, J.C., T. Takizawa, and J.D. Longstreth, 1989. *Human Health Environmental Effects Panel Report*, Pursuant to Article 6 of the Montreal Protocol on Substances that Deplete the Ozone Layer. United Nations Environment Programme (UNEP), Nairobi, Kenya.

Villach, 1985. *Report of the International Conference on the Assessment of the Role of Carbon Dioxide and of Other Greenhouse Gases in Climate Variations and Associated Impacts*. Report No. WMO-No. 661, International Council of Scientific Unions, United Nations Environment Programme and the World Meteorological Organization, 78 pp.

ATMOSPHERIC CHANGE AND TEMPERATURE-RELATED HEALTH EFFECTS

Edwin M. Kilbourne

Health Studies Branch
Division of Environmental Hazards and Health Effects
Center for Environmental Health and Injury Control
Centers for Disease Control
1600 Clifton Road
Atlanta, Georgia 30333

Introduction

Recent history is full of examples of short-sighted management of the environment and the consequent development of environmental conditions that endanger human health. Thus, I welcome the opportunity to be part of the farsighted effort in which we are engaged today. Seldom have so many people paid so much attention to the possible consequences 50 or 60 years hence that may occur as the result of present actions. While substantial uncertainty still exists regarding the nature and extent of the climate changes that may arise from anthropogenic changes in the composition of the atmosphere, the potentially devastating environmental impacts predicted by some investigators clearly warrant both our attention and the initiation of systematic efforts to investigate scientifically the scope of the possible health effects that may result from global climate change.[1]

Environmental Questions

Data developed by a number of investigators appear to show a trend toward increasing global temperature occurring over approximately the past 100 years. The appropriate interpretation of these data has been disputed and, since I personally lack the experience in atmospheric physics required to interpret these data, I will refrain from concluding on the basis of this information that significant global warming has either already occurred or is occurring. Nevertheless, data showing an apparent increase in average global temperature are in general agreement with forecasts derived from so-called "general circulation models" (GCMs) that attempt to model climatologic events on a global scale and which predict a rise in average global temperature of as many as 2 to 5 °C by the middle of the next century.[1] A number of dramatic environmental consequences appear likely if global warming of this magnitude occurs, including a rise in sea level, develop-

Published 1990 by Elsevier Science Publishing Co., Inc.
Global Atmospheric Change and Public Health
James C. White, Editor

ment of both new wetlands and areas stricken by drought, and an increase in the number and strength of some types of storms and other extreme weather events.[1] It therefore seems appropriate to attempt to determine the potential health impact of such dramatic changes. Answers to the following questions are particularly important: (1) In which individuals are adverse health effects particularly likely to occur? (2) Which geographic areas are most likely to be affected? (3) When and how frequently are particular health effects likely to be observed? (4) And finally, what actions might be effective in preventing adverse health effects from climate change?

I do not deal here with the problem of stratospheric ozone depletion, the consequent increase in biologically active ultraviolet radiation reaching the earth's surface, and the possible health consequences of those phenomena. Rather, I deal with the broad set of predicted climate changes often lumped under the single term "global warming."

Health Effects

General Questions

Potential health effects related to global warming can be categorized into at least one of four general categories: (1) direct physiological consequences of excess exposure to the heat (i.e. recognizable heat-related illness); (2) exacerbation by heat (or any other climatologic parameter related to global warming) of underlying chronic (noninfectious) medical conditions; (3) infectious diseases whose transmission is favored by climate change; and (4) health problems characteristic of refugee populations that may develop as people are displaced from their homes by the dramatic environmental events predicted to occur in association with global warming (for example, famine caused by drought and coastal flooding caused by a rise in sea level). There is, of course, substantial overlap among the third and fourth items in this list, since the health problems of special importance among refugees are largely infectious diseases.[2] My own area of interest and expertise relates to the first two items on the list, and my further discussion will center around direct heat-related illness and the exacerbation of certain chronic disease processes by heat.

Research in these areas has hardly begun but necessarily relates to several key underlying questions: Current problems in public health are already complex, so how is society to deal logically and effectively with future determinants of health and disease, around the predictions of which there is necessarily some margin of error? Epidemiology, the study of disease determinants in populations, generally deals with health events that have already occurred. Can epidemiologic methods be adapted so as to reliably predict patterns of disease occurrence that do not yet exist? Can the study of present day disease/exposure relationships be used to predict such patterns in the future? My own belief is that all of these questions can eventually be answered in the affirmative, but that much work lies ahead of us before we can predict future public health events with any precision.

Importance of "Average Temperature"

From the point of view of health effects, the anticipated extent of the changes generally predicted for the middle of the next century, an increase of average global temperature of from 2 to 5 °C, does not sound terribly threatening. To some, these numbers may sound positively pleasant, conjuring up visions of winters that are shorter and less harsh, as well as earlier springs. Nevertheless, the increase in summer heat could have devastating effects, at least in certain areas. From Figure 1, which shows annual heat-related deaths in the State of Missouri as a function of average summer temperature, it is evident that small changes in average temperature can be associated with a relatively large health impact. The average (daily maximum) temperature during summer of 1980, which saw approximately 300 heat-related deaths in Missouri, was only about 2 °C higher, on the average, than the summer of 1983, during which only about 50 heat-related deaths were reported.[3] Further, the correlation of higher average summer temperatures with increased numbers of deaths attributable to the heat is evident from the figure and suggests further adverse health consequences to come, should average summer temperatures rise.

Clearly, a number of factors may modify this dire scenario. Milder winters (particularly in a place like Missouri where they are often severe) may enhance people's capacity to physiologically acclimatize to hotter summers. Changes in architecture and building construction practices may favor greater heat tolerance in the population. People may learn to deal better with the heat by adapting their behavior, learning to spend more time in air-conditioned areas, a practice that may protect them from heat-related health effects to a substantial degree.[4]

Seasonal Trends in Mortality

As a point of reference, it may be useful to review the typical seasonal pattern of mortality in the United States and other temperate countries (Table 1). In general, when mortality from all causes is considered, the average daily death rate is substantially higher in winter than in summer months. The extent of this seasonal effect is surprisingly great. There are literally tens of thousands more deaths per month during winter than during summer months.[5] The complexity of this phenomenon becomes apparent when one considers that the current pattern has varied over time. Several decades ago, the months with the lowest mortality occurred in the late spring.[6] This apparent change may be the result of improved control and treatment of serious enteric illnesses (e.g. polio, infant diarrhea, and others) which have been important causes of morbidity and mortality during summer months.

The winter death excess is also an age-dependent phenomenon. It applies exclusively to the elderly and becomes more pronounced the older the age group. In fact, in persons less than 45 years of age, the seasonal trend in mortality is reversed, with increased mortality occurring during warmer months of the year.[7]

The increased mortality during winter months might lead one to the conclusion that an increase in average global temperature ought to promote health by decreasing this effect. However, certain data suggest that the winter death excess may not be strictly temperature-related. Unlike heat waves (see below), cold waves are not accompanied by such dramatic increases in local crude mortality rates. Moreover, in the United States, the winter death excess occurs in warm states as well as cold ones and is often of approximately the same magnitude in the warm as in the colder states.[7]

Heat Wave-Related Mortality

Heat waves, in contrast, are associated with clear-cut, abrupt increases in crude mortality. The 1980 heat wave caused substantial mortality increases in St. Louis and Kansas City, Missouri. At its peak, the number of daily deaths in St. Louis literally tripled as a result of the heat.[8] Such dramatic increases in numbers of deaths are far greater than those associated with the winter death excess.

The joint effects of the winter death excess and heat wave-associated mortality lead, in many U.S. cities, to the pattern of average number of daily deaths shown in Figure 2. At low maximum daily temperatures, which principally occur in wintertime, mortality is relatively high and falls progressively as the average daily maximum temperature increases. However, when maximum daily temperature increases above a certain point, the heat wave effect becomes noticeable and mortality increases. The heat wave effect is a clear result of increased temperature, whereas the high winter death rates, while associated with relatively low temperature, probably represent a complex set of seasonal factors. Thus, the mortality response that may occur as a result of global warming (increases in average temperature) are not so much of concern because of increases in the average temperature so much as the increased number and severity of heat waves (extreme events) expected under some climate change scenarios. While heat-sensitive cities such as St. Louis may adapt to the climatic change over the next 40-60 years, we do not know for certain that such adaptation will take place. If it does not, the results of increases in average temperatures and the resulting increases in heat waves could cause severe health consequences.

Clinical Aspects of Heat-Related Mortality

One's first impulse is to attribute excess deaths that occur during heat waves to heatstroke, the only highly fatal illness that is both caused by overexposure to the heat and is clearly attributable to heat exposure based on clinical examination of the patient. Heatstroke is a medical emergency and involves a dramatic clinical presentation. Heatstroke patients develop altered mental functioning that often rapidly evolves to coma. The skin is hot, and perspiration is often (but not universally) absent. Body temperature is 105 °F or greater. If the patient is not cooled rapidly, severe permanent neurological damage or death may occur.

An unexpectedly low proportion of deaths in heat wave episodes is made up by heatstroke deaths. The proportion of excess deaths attributed to heatstroke during heat waves runs from none to about 50%. Nevertheless, substantial excesses in mortality from cardiovascular disease, cerebrovascular disease (stroke), and other chronic disease categories are frequent.[9] Recent evidence suggests that heat stress may induce a mild hypercoagulable state.[10,11] Such a hypercoagulable state could precipitate formation of thrombi (blood clots) in the blood vessels of persons with some degree of underlying atherosclerotic vascular disease, a common finding in older persons. The result could be an increase in the number of coronary thromboses (heart attacks), accounting for the cardiovascular disease mortality excess during heat. Similarly, cerebral thrombosis or embolism could result in increased stroke deaths during a heat wave. This is an area that requires substantial further study. The sequences of events just described suggest that persons with underlying atherosclerosis may be at particularly high risk during the heat, an important point that requires further study because of its potential impact in focusing the preventive efforts of public health officials.

The Future

Given the clear-cut hazard posed by heat-related illnesses, including both heatstroke and heat-exacerbated chronic diseases, where should we go from here in our attempts to predict and control their likely impacts? This question has relevance whether or not significant global warming occurs, since heat-related illness is already a substantial public health problem in the United States.

A major difficulty with current efforts to prevent heat-related illness is the resource intensive nature of effective prevention programs (for example, the setting up of air-conditioned heat wave shelters), which generally are the responsibility of local health departments. Such programs are necessary during summers in which major heat waves occur but, since such heat waves do not occur every summer, resources dedicated to heat wave preparedness may be wasted, a cost that few local health departments can afford. Some sort of early warning system by which local health departments could gear up to defend against the effects of the heat would be of great help, but no such system is either employed on a widespread basis or of demonstrated utility.

The current availability of relatively reliable information on anticipated weather conditions several days in advance of their occurrence (weather forecasts) suggests the possibility of also predicting the extent of heat-related health effects that might occur as a result of the anticipated weather. These predictions would give public health officials as much as several days warning prior to the onset of dangerous meteorological conditions.

Our group at the Centers for Disease Control (CDC) has been undertaking research directed toward achieving this goal. Specifically we have attempted to model crude mortality in the City of St. Louis, Missouri, as a function of maximum daily temperature after adjustment for the effects of humidity by the Apparent Temperature (AT) method of Steadman.[12]

Our results, while preliminary, are promising. Using ten years of mortality and climatologic data, we constructed a model of mortality (adjusted for population changes over time) as a function of whether or not maximum daily AT exceeded three systematically chosen cutoff points: 38 °C, 40 °C, and 42 °C. Parameters for this "intervention model" (the interventions being dichotomous variables indicating whether each of the

AT cutoff points had been exceeded) were estimated using procedures described by Box and Jenkins.[13] Additional parameters were included in the regression model to deal with the problems posed by serial autocorrelation in the time series under study. The result was a model from which anticipated mortality was predicted relatively well from past values of mortality and AT (Figure 3).

An important caveat should be added here regarding the interpretation of models and forecasts such as those in the system just described. Figure 3 shows only a model's "prediction" of the observations that were actually used in its derivation. While the apparent close fit between predicted observed values is impressive, it really only shows that the model would have predicted past events well. The fit with past data does not necessarily guarantee that the model will work well in predicting the future. The development of forecast systems is necessarily an iterative process in which the deficiencies in existing models are identified as they fail to correctly predict real events. The identified deficiencies can then be remedied and the process repeated until a working predictive system is developed. We have a long way to go before we can confidently predict the extent of weather-related mortality from past mortality experience and meteorologic predictions. Nevertheless, the possibility is an exciting one and requires further vigorous pursuit.

There are additional problems involved in applying models designed to predict near- or intermediate-term phenomena to the prediction of health effects 40 to 60 years from now. The frequency of underlying health problems that can be exacerbated by heat could change dramatically, substantially decreasing the relevance of models derived from current health data. Moreover, the future societal adaptation (e.g. changes in urban structure, architecture, and health care delivery) to possible changes in climate several decades hence cannot be easily evaluated. Our lack of complete data will inevitably continue to limit the degree of certainty associated with the predictions we make regarding future events.

Conclusion

Study of possible global warming and its environmental consequences by atmospheric chemists, climatologists and scientists from other disciplines should be associated with parallel study by health scientists of the possible health effects associated with these environmental changes. The types of adverse health consequences that may occur need to be identified and their extent quantified to the best of our current ability. An important area of current endeavor is the development of models for the quantitative prediction of specific weather-related health effects. The development of such models and the systematic study of their deficiencies should eventually allow both fuller understanding of the relationship of climate to human health and a better grasp of the health consequences of any long-term changes in global climate.

NOTE: For correspondence regarding this manuscript, please contact Dr. Kilbourne at: Health Studies Branch, Division of Environmental Hazards and Health Effects (F-28), Center for Environmental Health and Injury Control, Centers for Disease Control, Atlanta, Georgia 30333 (Phone 404-488-4682; FAX 404-488-4308).

Table 1.   U.S. Deaths (All Causes) by Month, 1983–1987

Month	1983	1984	1985	1986	1987
January	181,957	183,684	201,669	194,905	196,512
February	168,261	169,385	184,323	178,546	170,579
March	178,959	181,554	182,908	190,248	186,971
April	169,748	169,810	169,073	172,719	175,488
May	167,375	169,810	168,585	172,391	175,098
June	158,115	162,206	161,062	163,740	167,548
July	164,905	161,833	164,916	170,204	172,202
August	160,302	160,897	162,748	166,234	170,290
September	155,776	158,533	161,753	163,316	165,068
October	166,612	168,186	172,517	172,801	180,965
November	166,645	170,516	170,305	174,650	175,900
December	183,535	184,069	189,519	188,630	189,721

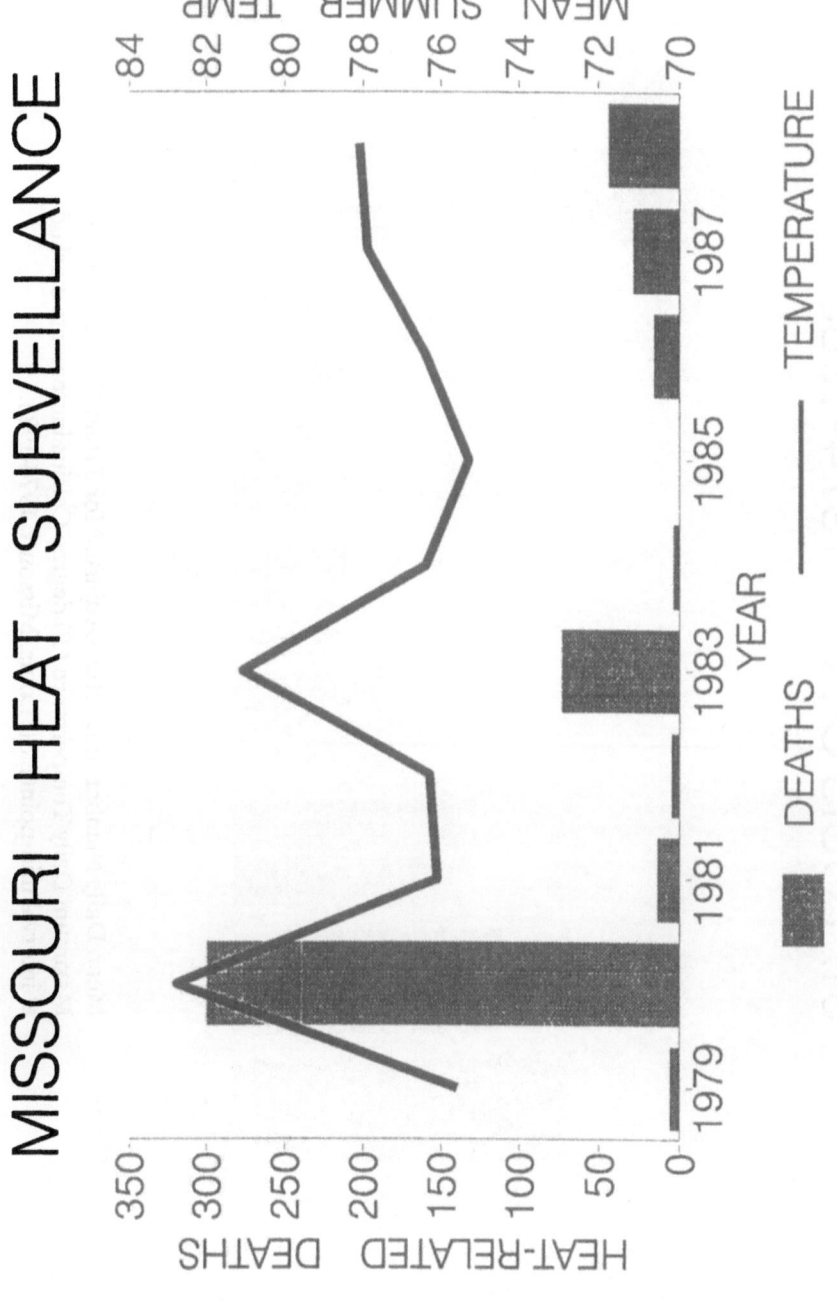

Figure 1.   Heat-Related Deaths and State Areally Weighted Temperature (Mean
June.-August) by year, 1979-1988, Missouri, United States.

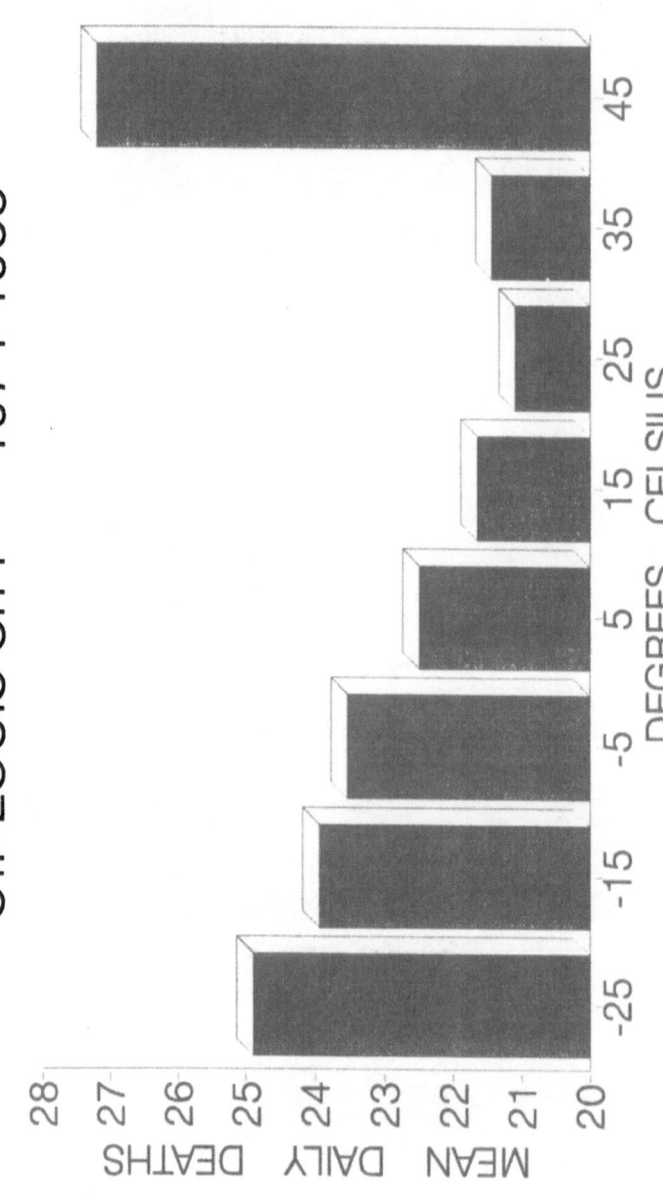

Figure 2.   Mean Daily Number of Deaths (Adjusted for Trend Over Time), by Maximum Daily Temperature (10 degree Celsius increments marked at interval mid-point), St. Louis, Missouri, 1974-1983.

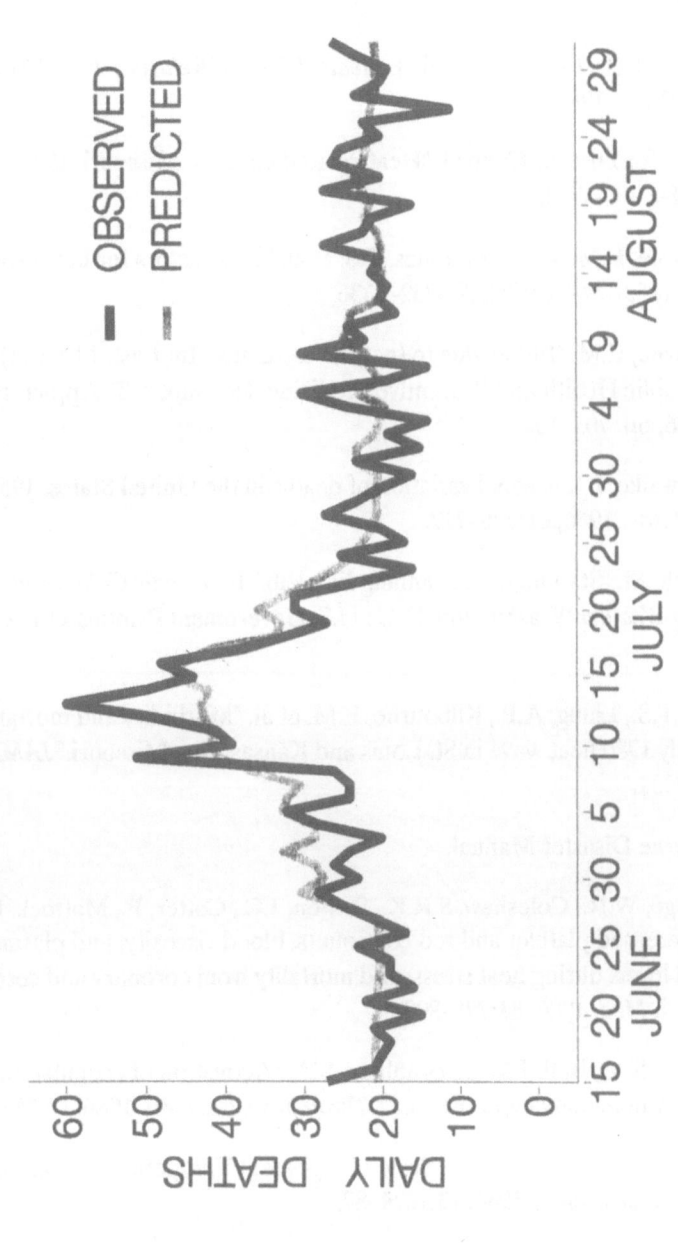

PREDICTED & OBSERVED MORTALITY
APPARENT TEMPERATURE MODEL - ST. LOUIS - 1980

Figure 3. Predicted and Observed Deaths (Adjusted for Time Trend) by Day
During the 1980 Heat Wave, St. Louis, Missouri.

References

1. Schneider, S.H. "The greenhouse effect: Science and policy." *Science* 1989; **243**:771–81.

2. Sandler, R.H., Jones, T.C. (eds.). *Medical Care of Refugees.* New York: Oxford University Press, 1987.

3. Centers for Disease Control. "Heat-related deaths – Missouri, 1979–1988." *MMWR* 1989; **38**:437–9.

4. Kilbourne, E.M., Choi, K., Jones, T.S. et al. "Risk factors for heatstroke. A case-control study." *JAMA* 1982; **247**:3332–3336.

5. Kilbourne, E.M. "Illness due to thermal extremes." In: Last, J.M. (ed). Maxcy-Rosenau Public Health and Preventive Medicine. Norwalk, CT: Appleton-Century-Crofts, 1986, pp. 703–10.

6. Rosenwaike, I. "Seasonal variation of deaths in the United States, 1951–1960." *J. Am. Stat. Assoc.* 1966; **61**:706–719.

7. Feinlieb, M. "Statement of Manning Feinleib." In: *Deadly Cold: Health Hazards due to Cold Weather.* Washington, D.C.: U.S. Government Printing Office, 1984, pp. 85–125.

8. Jones, T.S., Liang, A.P., Kilbourne, E.M. et al. "Morbidity and mortality associated with the July 1980 heat wave in St. Louis and Kansas City, Missouri." *JAMA* 1982; **247**:3327–3331.

9. Kilbourne Disaster Manual.

10. Keatinge, W.R., Coleshaw, S.R.K., Easton, J.C., Cotter, F., Mattock, M.B., Chelliah, R. "Increased platelet and red cell counts, blood viscosity, and plasma cholesterol levels during heat stress, and mortality from coronary and cerebral thrombosis." *Am. J. Med.* 1986; **81**:795–800.

11. Strother, S.V., Bull, J.M.C., Branham, S.A. "Activation of coagulation during therapeutic whole body hyperthermia." *Thrombosis Research* 1986; **43**:353–60.

12. Steadman, R.G. "A Universal Scale of Apparent Temperature." *Journal of Climate and Applied Meteorology* 1984; **23**:1674–87.

13. Box, G.E.P., Jenkins, G.M. *Time Series Analysis: Forecasting and Control.* Oakland, CA: Holden-Day, 1976.

GLOWNET: AN INFORMATION CLEARINGHOUSE NETWORK CONCEPT

E. Joseph Bangiolo

National Institute of Allergy and Infectious Diseases
900 Rockville Pike
Bethesda, MD 20892

Abstract

Studies of global warming and myriad allied fields are generating volumes of information. More and significant information diversity may be expected. Means are needed to deal with diversity and growth in each of many venues: the research, education and public arenas. To help synthesize information handling resources, GLOWNET, a clearinghouse network model, is proposed for consideration.

Introduction

Emerging scientific and political consensus is that global environmental status can be and should be researched and tracked. There are great threats to health, human security and survival (especially economic survival) if true global climatic change occurs more quickly than human adaptability. This would happen if there is a rapid onset (say, less than 50-100 years) of extensive global warming. Technology (satellites, computers, mathematics and communications) is available to generate successful research on global status and changes. There should be emphasis on information resource development in specific areas: research, education and communication/dissemination.

Three foreseeable situations exist for information resources development: (1) it may lead and catalyze science and social/political action; (2) it may develop concurrently; or (3) it may lag the field (as is typical and perhaps the expected case). Timeliness and leadership directions vis-à-vis information resource development may be critical to effectiveness in dealing with the problems and issues of global warming.

There is a significant need to assess the existing information resources on global warming and environmental status changes and the potential contribution of contemporary or advanced systems of information science and technology.

Global Atmospheric Change and Public Health
James C. White, Editor

Growth and diversity of information on global warming and environmental status is emerging as a natural consequence of technologies. As a result, there is a need to use information sciences/technologies to deal with research, communications and information dissemination.

Volume and diversity of information impose significant information management problems. To be useful, for example, information sciences must make contributions in the area of development of information tools such as vocabulary, linguistic techniques, thesauri, etc. Other areas include automated indexing, storage/retrieval systems, and dissemination.

New Language/Old Languages

The language problem is significant. New terms are needed. Researchers must cope with the international nature of the problem. Language, specifically terms and phrases, often confuses as well as enlightens. Jargon must be avoided, but meaningful technical expressions can communicate and express new situations and concepts.

"Terminologists" are needed for coining new terms. One can imagine words such as megaenvironmental, macroclimates and supertrend (or equivalents) coming into wide use as they are defined and popularized.

Specific needs include development and use of language for describing massive changes that occur with glacial speed and whose only manifestation is diversely collected data with microchanges in timing, frequency and spatial values. New linguistic elements may be needed to interpret and express the rapid changes in previously "unchanging" or slowly changing environmental factors. As the language continues to develop, information science can be applied to foster effective management of the needed information and data resources.

Specific tools such as thesauri and indexing techniques will be derived from existing tools, and perhaps new formats or approaches will supplant even our most useful present methods. This is an exciting challenge in itself. The key point is to identify the words to fit the music so that world science can keep track and communicate effectively.

For a considerable portion of mankind's existence, the flat world model was an effective and useful model. New terms, descriptions and language are needed for the global world. To use any model, terms are key. In a model that expresses a warming climate, change becomes significant only if the speed of the change is increased and recognized with descriptions and measurement. With time and technology, the flat world model of the globe was found to be flawed. Farming is just fine with the flat world model, but long-distance navigation, communications and global commerce is not. Part of the difficulty in understanding and acting on the global model was a failure of the information sciences of the time. There were no good terms that were understandable.

Translation of world-wide research can be expected as a need. International cooperation and communications will be of particular importance. This ancient problem merely enters a new arena with gross global environmental information needs. It is expected that methods that now cope with broad fields such as world-wide biomedical research, economic/monetary data and industrial technology would find ample application.

Storage/Retrieval Technology

Technologists must develop consistent means for dealing with the usual problems of effective storage into media and structures that permit efficient retrieval and use. These areas are undergoing a pure revolution due to computer techniques for manipulation of new media. "Gigabyte" storage in accessible media is a reality. Computer speeds and software permit highly effective retrieval even from the technologist's desktop.

A special challenge is arranging intellectual schemes for effective identification, and development of methods for organizing the stored information with specificity, accuracy and yields.

Dissemination

Dissemination is a special need and a significant challenge in the climatic sciences. "Everyone talks about the weather, but no one does anything about it." If some of our concerns about environmental science are realistic, the new epigram will be, "Everyone knew the climate was changing, but who will take the blame for ignorance in acting on this knowledge?"

Dissemination and diffusion of information is a powerful influence. Unfortunately, this influence may produce false leads, ineffective policy, stunning inaction and other undesirable effects, in addition to needed positive pursuits, actions and attitudes.

Earlier, popular obeisance was paid to the journalistic aphorism, "I only know what I read in the newspaper." Updating this, even many influential people may think, "If it isn't on television, it does not exist." Television is a powerful consumer of information resources. Appropriate quantity and quality of this resource is a need.

Ample opportunity exists for the information community to deal with emerging information and data if they are sensitized. Large-scale changes with small incremental indications may make the world end "not with a bang but a whimper." However, "whimpers" do not make good "sound bytes" for television.

What is "effective" information? This is information that enlightens, sensitizes, alerts and produces the foundation of action. It is information that describes the past with needed accuracy, the present with broad perspective, and shines a beam on the plausible future.

A Synthesis

Assessment of information resources is a priority need. Thus what constitutes a program of assessment? To begin, top-level examination of essential leadership/policy issues is a need. This can be accomplished by a continuing process bringing together researchers and information technologists in an appropriate forum. Information technologists must make a concentrated effort to confront the problem. Its scope (complexity, diversity, volume and growth) is exquisite. A deliberate effort requires multiple analysis of needs, priorities and resources.

The present conference is a significant step. This could lead to the formulation of a program based on policies such as:

(1) Global warming is an international problem which requires a concerted awareness of the information issues by leaders in science, industry and government.

(2) The information can be organized into components. These include formulation, information processes, and resources and activities with a focus on using the information sciences and technology.

(3) Top-level assessment of the status of information processes, resources and activities should be conducted by experts in science, government and the information sciences.

(4) The assessment phases of the programmatic examination of the problem will allow the setting of goals, objectives and approaches to deal with the information issues, including public and professional education.

(5) If there is an early consensus on goals and objectives, one means of implementation is an enhanced clearinghouse network.

Next Steps

What is needed next? The programmatic establishment, expanding and intensifying of a clearinghouse network in conjunction with the Air Resources Information Clearinghouse is perhaps a feasible step. Areas for discussion include: defining the scope, organizing information exchange on policy, appropriately judging the magnitude of the problem, and directing assessment processes.

Numerous clearinghouses, information centers and information resources exist in government and nongovernment sectors world-wide. A process of involving increased liaison and other contact is recommended. More communications will allow for the sharing of approaches and mutually beneficial activities.

Two steps can continue the process of cooperation and coordination to enhance a clearinghouse network:

- Assessment and policy development by the leadership community: Leadership by researchers, technologists and information experts can assure the continuing and timely assessment of information on global warming, macroclimatic changes, economic effects, health effects and other megatrends.

- Formulation of new patterns and approaches: New intellectual combinations and patterns can contribute to the fund of information, its organization, processing, dissemination and diffusion. This assessment is on-going today as evidenced by the growing literature on issues of global warming and the excellent presentations at this conference.

Furthermore, the public is being alerted to the issue by communications media. The whole subject matter has an intrinsic interest. This will lead to wide use by researchers, journalists and others in publications, articles, interviews, workshops and in other popular and public forums.

"GLOWNET" Clearinghouse Network: A Model of Information Resources Consortia

The following are proposed steps for a clearinghouse network:

- Formulation of information policies for the survey and design of a clearinghouse network (GLOWNET). This would include assessment of the existing situation and collecting information on the roles and interests of existing clearinghouses/information centers and other information resources.

- Structuring information topologies including selecting goals and objectives, setting priorities, and organizing appropriate resources and implementation projects. Information topologies to be selected may come from lists such as the following:

   - Top-level policy and program information — Information on organizing and funding priority programs.

   - The literature — Abstracts, indexes and full texts. Accessible from advanced media such as on-line databases, telecommunications, workshops, conferences, etc., as well as traditional means.

   - The data as handled, processed, winnowed and refined by researchers — GLOWNET would assist in the formation of Special Interest Groups (SIGs),

Information Exchange Groups (IEGs), and general intensification, broadening and expansion of collegial contacts, communications and consortia.

Implementing the Program

What is proposed is a multi-functional clearinghouse network that would link communications, research information and program information with users, audiences and publics via a voluntary and structured consortium (a network), information resources and clearinghouses. The concept is entitled "GLOWNET," a name that embodies GLObal Warming as a motif.

The form of GLOWNET is ephemeral at present. What is certain, as the emerging research information volume and complexity continues, is that means for its management and beneficial exploitation will be found.

Potential GLOWNET functions include:

- Collection of literature from journals and other sources
- Collection of program descriptions
- Access to major databases (thus forming the clearinghouse network)
- Analysis of information sources, gaps and trends
- Sponsorship of workshops, meetings and conferences
- Development of an exhibits program on global warming information
- A "hot line" or 800 telephone service
- Development of internship programs
- Scholarships and awards programs for contributions to the field
- Sponsorship of Special Interest Groups and Information Exchange Groups
- Organization of an on-line database (with selected projects and abstracts)
- A specialized newsletter (electronic and print formats)
- Publication of monographs, articles, bibliographies and other print media
- A schools program with science education content

Potential Role of ARIC

Under the auspices of the Air Resources Information Clearinghouse, GLOWNET would be an expanded, intensified and broadened component.

This effort would capitalize on existing information resources in the U.S. Government sector including those of the National Technical Information Service and others of the U.S. Department of Commerce, the National Library of Medicine (NLM) Toxline, the Toxnet systems (RTECS, HSDB, CCRIS, etc.), systems of NOAA, EPA, NSF, USDA and numerous other agencies concerned with the underlying sciences.

GLOWNET would call on the private sector as well, the Academy of Sciences and numerous others in the academic and information industries (vendors of large telecommunication database systems).

It is then left as a challenge to information technologists, with interests and contacts in the broadening fields mentioned, to lead, to maintain or to follow with the needed mechanisms, implementations and operations. The question of funding and management are also waiting, but we hope will not get in the way so as to keep GLOWNET from being global or warmly received.

NEDRES: "YELLOW PAGES" DIRECTORY TO ENVIRONMENTAL DATA

Gerald S. Barton

National Environmental Data Referral Service
National Oceanographic Data Center
National Oceanic and Atmospheric Administration
1825 Connecticut Avenue NW
Washington DC 20235

Introduction

The need for directories of data and information is not new. Indexes, catalogs, and directories have been assembled and published in printed format for many years. Efforts to present this information in a computer-accessible format recently have been increasing at a dramatic pace. Large scientific programs are recognizing that the key to scientific research is the ability to locate data that have already been collected. On-line computer directories help to service this need.

The National Oceanic and Atmospheric Administration (NOAA) began work on a computer directory to environmental data in the early 1970's with the development of the Environmental Data Index (ENDEX). The system used an early Data Base Management System (DBMS) on an IBM computer. It did not have on-line capabilities and clients were serviced by NOAA researchers using batch computer runs.

The National Environmental Data Referral Service (NEDRES) is the yellow pages directory for locating environmental data in the United States. NEDRES was developed using the ENDEX description format as a model. NOAA provides the service as part of its responsibility to archive and document environmental data. The database is accessed via a local telephone call using a computer terminal or personal computer. A search is made using any word desired by the user, and the results can be listed on the user's terminal or printed at the computer site.

Data from all sources are described in NEDRES. It is not limited to NOAA data, but describes data from federal, state and local government organizations, private companies, academic institutions, and individual researchers. Examples of the diversity of sources of data in NEDRES include several State University of New York locations, Virginia State Water Control Board, and Climate Assessment Technology, Inc. in Houston, Texas.

Published 1990 by Elsevier Science Publishing Co., Inc.
Global Atmospheric Change and Public Health
James C. White, Editor

The NEDRES database identifies the existence, location, characteristics and availability of environmental data. It is a publicly available service which allows easy access to data information that could be useful for problem solving. NEDRES describes the data and directs the user to the holder of the data, but the data are not available from NEDRES.

Climate, meteorology, satellite remote sensing, pollution, fisheries and oceanography are examples of the major categories of data described in the database. The scientist can search using any desired descriptor to locate data of interest. For example, a marine scientist might use bathymetry and geology as descriptors for a search. This tool is a valuable contribution to research since it directs the user to a wide variety of data sources.

There are other ways that NEDRES can be used. Special publications can be automatically generated on the computer. For example, "North American Climate Data Catalog Part I" (NOAA 1984), "North American Climate Data Catalog Part II" (NOAA 1985a), "Specialized Data Catalog: Chesapeake Bay and Wetlands" (NOAA 1985b), "Specialized Data Catalog: Coastal and Estuarine Waters of California, Oregon, and Washington" (NOAA 1985c), "Satellite Remote Sensing of the Marine Environment: Literature and Data Sources" (Barton et al., 1986), and "Chesapeake Bay Environmental Data Directory" (Jacobs et al., 1987) were produced as a result of tailored searches of the database.

The NEDRES Database

NEDRES uses the facilities of a commercial computer company, BRS Information Technologies, located in Chicago, Illinois. The system is available from any location in the United States or the world using commercial telecommunications networks such as Telenet. Users may obtain a password to use NEDRES by contacting BRS. The user also has access to all of the other databases available on BRS, such as the PHD Dissertation database. Contact the NEDRES Office on 202-673-5548, or BRS Customer Service on 800-289-4BRS. BRS has several options for using the system; for example, nighttime usage costs are about one-third of prime time usage.

The NEDRES database management system combines sophisticated and flexible searching techniques with an easy-to-use menu driven system. The user can sign on to the system, enter a search query, obtain results, and sign off with a minimum charge. The database messages and commands are simple, consisting of English language words instead of numeric or code representations.

A search of the NEDRES database provides a complete description of available data sources that satisfy the search specifications. The resulting information describes the data in sufficient detail allowing the user to decide whether to contact the data holder for specific details or to arrange to acquire the data. Separate fields describe the Title (TI), Abstract (AB), Data Collection (DC), Data Processing and Quality Control (DD),

Period (PE) and Length of Record (LR), Geographic Location (GE, GC, GL), Description of Parameters (PA), Contact to obtain the data (C0), Availability Characteristics (AV), Principal Investigators (PI), Program or Project (PR), Processing Organization (PO), Related Publications (PU), Accession Number (AN), Accession Date (DT), and Discipline Codes (CC).

The database documents several types of environmental information descriptions. It includes:

1. data centers, programs and organizations;
2. data files not in published forms;
3. data serial publications;
4. published data sets;
5. atlases or published data in graphic or analog form;
6. publication containing extensive compilations, analyses or applications of data;
7. manual, user guide, or documentation of a data set;
8. data catalog, inventory, or bibliography.

The variety of data in NEDRES serves users from many fields, allowing a multidisciplinary search for data information. In most other systems the data are for a specific discipline and several data catalogs might have to be searched to satisfy a user's needs. With NEDRES the user can find data from different disciplines and combine the searches to isolate the data desired. As an example, the user may need to locate physical and biological oceanographic data, climatic data, and marine geological data for a study in the Gulf of Mexico. By using the search capabilities of the NEDRES database, all of the data meeting the criteria would be retrieved, and information describing the data would be immediately available to the researcher.

There are currently over 22,200 data descriptions in the database. The coverage is global and most of the data are held in the United States.

Using NEDRES To Find Environmental Data

NEDRES provides the decision making and research communities with a tool for finding the existence and availability of data information related to the environment. The scientist may search for data using any search criteria and relationships desired. For example, a search could find all records in the NEDRES database which describe toxic data for the Chesapeake Bay.

There are no caveats on the way NEDRES is used. For example, there is no need for the user to refer to a list of key-words in order to perform a search. The database indexes every word in a data description except for a list of stopper words such as: the, and, not, to, etc. To search for a word of interest, the user merely types the word, such as "toxic," enters a carriage return, and the system responds with the result of the search.

The print image of the session would look like this:

> 1_: toxic
> RESULT 2536

This means that there are 2536 data descriptions in the NEDRES database that have the word toxic somewhere in the record.

There are powerful search facilities available on the system. Searches can be combined using Boolean operators, for example:

> 1_: wind and temperature

which searches for wind and temperature in the same record. Search words can be limited to specific sub-record areas such as the Geographic Area. Words that occur adjacent to each other can be specified as in:

> 1_: marine adj geology

which searches for marine geology in the record. Another useful search tool is the ability to define that two words are located in the same paragraph as in:

> 1_: estuary same pollution

which searches for those sub-records (such as the abstract) which have both estuary and pollution mentioned.

Table 1 is a sample of titles of data descriptions selected from the database. It shows the broad scope of environmental data in NEDRES.

A complete description from the database is given in Table 2. The search parameters were:

> 1_: pollution
> RESULT 630
>
> 2_: Pennsylvania
> RESULT 619
>
> 3_: 1 and 2
> RESULT 40

which resulted in 40 records satisfying the search criteria. The search was done in three stages: "1." resulted in 630 records containing the word pollution, "2." resulted in 619 records containing Pennsylvania, and "3." combined pollution and Pennsylvania which resulted in 40 records containing pollution in Pennsylvania. The record selected for Table 2 is "Environmental radionuclide concentrations in the vicinity of the Peach Bottom Atomic Power Station, Pennsylvania." Note the comprehensive information given in the PA (Parameter) and DE (Descriptor) fields. The DE field contains both chemical terms and biological terms with corresponding taxonomic codes.

The use of the NEDRES database is described in the NEDS Database Guide which is available from BRS Information Technologies, 800-289-4BRS.

Use of NEDRES by the Medical and
Public Health Communities

The BRS Information Technologies system has a diverse user community. Users from many areas of the world use the system via telecommunication networks. The research and information communities are the major users, but about one-quarter of the users are from the medical and public health communities.

BRS caters to the medical communities in two ways. It provides access to about 20 different medical databases and offers users the COLLEAGUE service. COLLEAGUE allows users to search for colleagues who may be using the system, provides mail message service, and provides easy access to databases using features of the menu system.

A sample of NEDRES users from the medical and public health communities is shown in Table 3. Users are not merely from research institutions such as the National Institute of Health or the Duke University Medical Center, but range to small users such as private MD's and the Pediatric Academy Association in Ohio.

Because NEDRES is housed on the BRS system which has many medical and public health users, NEDRES provides the health community with an interface for finding environmental information which may be critical to public health studies in a global change environment.

A Developing National and International
Data Directory System

During the last two years, there has been a growing effort to develop environmental data directory systems by a number of organizations. The National Aeronautics and Space Administration developed the NASA Master Directory to describe NASA data sets and data systems. This effort has grown through involvement of NOAA and the U.S. Geological Survey. The NASA Master Directory has been used by the Interagency Working Group on Data Management for Global Change (IWGDMGC) as a prototype for global change data in the U.S.

Copies of the NASA Master Directory have been or will be installed in other locations, including NOAA, the European Space Agency in Italy, a facility in Japan, and the United Nations Environmental Program in Geneva, Switzerland, and Sioux Falls, South Dakota. The NOAA Earth Systems Data Directory will document NOAA data holdings. These systems, along with NEDRES and U.S. Geological Survey directories, form an international system of data directories.

As part of its directory effort, NASA staff developed the Directory Interchange Format (DIF), which provides standard definitions and formats for directory information. The DIF is documented in the "Directory Interchange Format Manual" (NASA 1989). The DIF Manual gives the syntax for each field in a data description record, gives the allowable entries on controlled fields, and gives sample data descriptions in Directory Interchange Format.

By using the DIF, directories can exchange data descriptions. The DIF will soon be proposed as an international standard for describing data in directories. This standard will provide a common link for existing and new directory systems.

Summary

NEDRES offers the researcher a powerful tool for identifying the existence, location, characteristics, and availability conditions of environmental data. Using it allows the scientist to find sources of data which meet any search criteria desired. The gathering of data and use of data is a means of promoting cooperation between scientists, researchers, and managers on both national and international levels. Use of NEDRES provides a unique way to locate interdisciplinary data which can be used by decision makers to solve complex problems affecting the environment of the nation and the world.

Because NEDRES is housed on the BRS system which has many medical and public health users, NEDRES provides the health community with an interface for finding environmental information which may be critical to public health studies in a global change environment.

A national and international system of directories is being developed to serve various user communities. The NASA Directory Interchange Format is becoming a standard for the exchange of data descriptions between different directories.

Table 1.   Titles Selected from the NEDRES Data Base

Atmospheric carbon dioxide in Hawaii.

Characteristics and extent of biological abnormalities within Puget Sound and identification and distribution of chemical contaminants.

Arctic Sea ice concentration grid.

SIO SE Asia sediment data file.

Geochemistry of polychlorinated biphenyls in the Hudson River.

Gulf of Mexico and South Atlantic Outer Continental Shelf study on the distribution
and abundance of endangered and vulnerable mammals, birds, and turtles, FY 80
(29096).

Toxicants in the Chesapeake Bay sediment – TOXIC.SEDIMENT,DIR.

Mississippi, Alabama, Florida (MAFLA) environmental monitoring program, FY 75
(CT5-30).

Fish distribution – MESA New York Bight atlas monograph 15.

Effects of brine pollution on aquatic organisms in three Michigan streams.

Trace metals in marine biota and sediments collected from offshore waters of the
New York Bight.

Cluster analysis of fish in a portion of the upper Potomac River.

A field evaluation of the effects of heated discharges on fish distribution, Virginia
power plant.

Experimental oil spill in Cub Creek, Virginia.

The Missouri lead study.

Investigation of the pollution of the St. Louis River below the junction of the Little
Swan, of the St. Louis Bay, and of Lake Superior adjacent to the cities of Duluth
and Superior (1929).

Ecological survey – Virginia Power's Portsmouth Power Station.

Environmental radionuclide concentrations in the vicinity of the Peach Bottom
Atomic Power Station, Pennsylvania.

Radiological environmental monitoring program, Calvert Cliffs Nuclear Power Plant,
Maryland.

Table 2. NEDRES Description of an Environmental Data Set

TI Environmental radionuclide concentrations in the vicinity of the
   Peach Bottom Atomic Power Station, Pennsylvania.
AB   The Maryland Department of Natural Resources, Power Plant Siting
   Program conducts radiological analyses of environmental samples
   collected from the Susquehanna River and Upper Chesapeake Bay.
   These studies were initiated in 1979 to determine the
   radioecological impact of the Three Mile Island Nuclear Station and
   the Peach Bottom Atomic Power Station on the Maryland environment.
   Collections of sediments, finfish, shellfish, aquatic mammals,
   aquatic vegetation, and waterfowl are analyzed by high-resolution
   gamma spectroscopy to determine gamma-emitting radionuclide
   concentrations. Analytical results and interpretations of studies
   are included in the reports. (ORNL).
DC Observing station type: Land, Ship.
   Instrumentation: Seine, hook and line, traps, mechanical box grab.
   Data collection type: Point location.
PE Earliest date: Yr 1979; Mo 00; Da 00; Hr 00
   Latest date: to present.
LR 7 years.
GE North Atlantic Ocean, Coast, Mid-Atlantic Bight, Chesapeake Bay,
   Susquehanna River, USA, Maryland, Harford County; Cecil County;
   Kent County; Baltimore County. North America, USA, Susquehanna
   River, Pennsylvania, York County, Delta, Peach Bottom Atomic Power
   Station; Lancaster County; Cumberland County; Dauphin County.
GC US24025. US24015. US24029. US24005. US4213318800. US42071.
   US42041. US42043. ER2214. ER2320. ER0030. WR020600. WR020503.
GL   SE Corner of Area   NW Corner of Area
   Latitude   Longitude   Latitude   Longitude
   N391500   W0754500   N401600   W0765200.
PA radioactive isotopes; Variable: concentration; Substrate:
       waterfowl; Part: flesh, heart, liver, kidney, gut,
       gizzard; Medium: atmosphere; Method: gamma
       spectroscopy; Units: picocuries per wet kilogram;
       Data type: observed.
   radioactive isotopes; Variable: concentration; Substrate:
       mammals; Part: flesh, gut, liver; Medium: earth
       surface, riverine, estuarine; Method: gamma
       spectroscopy; Units: picocuries per wet kilogram;
       Data type: observed.
   radioactive isotopes; Variable: concentration; Substrate:
       fish; Part: whole, flesh, gut; Medium: estuarine,
       riverine; Method: gamma spectroscopy; Units:
       picocuries per wet kilogram; Data type: observed.
   radioactive isotopes; Variable: concentration; Substrate:
       invertebrates; Part: flesh, meat, shell, gut;
       Medium: estuarine, riverine; Method: gamma
       spectroscopy; Units: picocuries per wet kilogram;
       Data type: observed.
   radioactive isotopes; Variable: concentration; Substrate:
       aquatic plants; Medium: estuarine, riverine; Method:

gamma spectroscopy; Units: picocuries per wet kilogram;
          Data type: observed.
     radioactive isotopes; Variable: concentration; Substrate:
          sediments; Medium: estuarine, riverine; Method:
          gamma spectroscopy; Units: picocuries per dry kilogram;
          Data type: observed.
DE Cesium-134. Cesium-137. Iodine-131. Zinc-65. Cobalt-60.
     Chromium-51. Manganese-54. Cadmium-109. Cobalt-58.
     Strontium-89. Strontium-90. Barium-140. Lanthanum-140.
     Silver-110. Potassium-40. Beryllium-7.
     Anthophyta I, Haloragaceae, Myriophyllum; water milfoil; TX32,
     TX324901; TX32490101.
     Anthophyta II, Hydrocharitaceae, Potamogetonaceae, Vallisneria
     americana, Potamogeton perfoliatus; wild celery, redheaded
     grass; TX33, TX330501, TX330605, TX33050103, TX3306050106.
     Mollusca, Bivalvia, Pelecypoda; mussels, clams; TX55. Chordata,
     Percidae, Centrarchidae, Cyprinidae, Ictaluridae, Clupeidae,
     Percichthyidae, Atherinidae, Stizostedion vitreum, Micropterus
     dolomieui, Cyprinus carpio, Ictalurus punctatus, Ambloplites
     rupestris, Pomoxis, Lepomis, Notropis, Dorosoma cepedianum, Alosa
     pseudoharengus, Notemigonus crysoleucas, Morone americana, Morone
     saxatilis; walleyes, smallmouth bass, carp, channel catfish, rock
     bass, crappies, sunfish, shiners, gizzard shads, alewife, golden
     shiners, white perch, striped bass, silversides, menhaden;
     TX883520, TX883516, TX877601, TX877702, TX874701, TX883572,
     TX880502, TX88351607, TX88351605, TX87760111, TX8835200401.
     Chordata, Anatidae, Muridae, Procyonidae, Mustelidae, Mergus
     serrator, Ondatra zibethicus, Procyon lotor, Lontra canadensis;
     mergansers, muskrats, raccoons, otters; TX911201, TX921602,
     TX922003, TX922002, TX9112012102, TX9216010401, TX9220030101,
     TX9220020201.
     Environmental monitoring. Water pollution.
CO Thomas E Magette
     301-269-2261
     Maryland Department of Natural Resources, Power Plant Siting
          Program
     Tawes State Office Building
     Annapolis, MD 21401   USA
AV The report is available from the Maryland Department of Natural
     Resources, Power Plant Siting Program. Data sheets are currently
     being transferred to a computer data base.
PU McLean, R. I.; Magette, T. E.; and Zobel, S. G. 1983. Environmental
     radionuclide concentrations in the vicinity of the Peach Bottom
     Atomic Power Station: 1979-1980. Maryland Power Plant Siting
     Program, PPSP-R-5, 29 pp.
AN 002392.
DT 19851231.
CC CC22 CC23 CC24 CC42 CC43 CC44 CC63 CC64 CC90 RT1 RT5
     CSUS-MARYLDEPARNATURRESOU.

Table 3. NEDRES Users from the Medical and Health Communities

Psychiatric Center, New York
Medical Center, Chicago, Illinois
Wakayama Medical College Library, Japan
MD in Vermont Department of Health
A. Epstein College Medicine, New York
MD in Massachusetts General Hospital
MD in Denton Texas
Upjohn Company, Michigan
USC School of Pharmacy, California
Duke University Medical Center
MD from Vancouver, Canada

References

National Oceanic and Atmospheric Administration, *North American Climatic Data Catalog Part I*, NEDRES Program Office, Washington D. C. 20235, 614 pp., 1984.

National Oceanic and Atmospheric Administration, *North American Climatic Data Catalog Part II*, NEDRES Program Office, Washington D. C. 20235, 614 pp., 1985a.

National Oceanic and Atmospheric Administration, *Specialized Data Catalog: Chesapeake Bay and Adjacent Wetlands*, NEDRES Program Office, Washington D. C. 20235, 349 pp., 1985b.

National Oceanic and Atmospheric Administration, *Specialized Data Catalog: Coastal and Estuarine Waters of California, Oregon, and Washington*, NEDRES Program Office, Washington D. C. 20235, 105 pp., 1985c.

Barton, G., E. Roberts, E. Riccio, and L. Stackpole, *Satellite Remote Sensing of the Marine Environment: Literature and Data Sources*, NEDRES Program Office, Washington D. C. 20235, 253 pp., 1986.

Jacobs, D., D. Haberman, D. Smith, D. Swartz, E. Sigel, A. Adams, *Chesapeake Bay Environmental Data Directory, Maryland Sea Grant*, University of Maryland, College Park, Maryland, 20742, 876 pp., 1987.

National Aeronautics and Space Administration, *Directory Interchange Format Manual*, NSSDC/WDC-A R&S 89-24, Goddard Space Flight Center, Greenbelt, Maryland, 20771, 89 pp., July 9, 1989.

A MODEL PROGRAM FOR PHYSICIAN EDUCATION IN ENVIRONMENTAL MEDICINE

Donna L. Orti and Max R. Lum

Public Health Service
Agency for Toxic Substances and Disease Registry
Chamblee 538, 600 Clifton Road
Atlanta, GA 30333

ABSTRACT

Studies indicate that the physician is the most trusted source of information in the community; however, in regard to environmental issues, physicians feel that they lack information and resources. Before primary care physicians can assume a more active role in environmental medicine, they will require readily available, accurate information. As an agency involved in promoting education about environmental medicine, we have been concerned about how best to present information to physicians. How can we make the issues of environmental medicine compete for a physician's time, especially when the occurrence of environmental illness is perceived to be infrequent? The Agency for Toxic Substances and Disease Registry (ATSDR) has prepared materials for the primary care physician about environmental medicine. The initial evaluation of the materials, *Case Studies in Environmental Medicine*, indicates that they are being well received and meet the needs of the physician. This model for educating physicians on the medical concerns of hazardous substances in the environment can be used to inform physicians of other environmental issues, such as global atmospheric change. Physicians educated in environmental medicine will be better able to respond to the questions of their patients, thereby making the transfer of reliable information more complete.

INTRODUCTION

During this meeting our discussions have focused on the present and future condition of our fragile world. We have heard about the impact of greenhouse gases and atmospheric warming on infectious diseases, food production, and human nutrition. We have learned about the depletion of the ozone layer and the surprising but shocking development of the ozone hole over Antarctica, and the effect of increased ultraviolet light on the spread of infectious diseases, the suppression of the immune system, the

Published 1990 by Elsevier Science Publishing Co., Inc.
Global Atmospheric Change and Public Health
James C. White, Editor

development of cataracts, and the increased incidence of skin cancer. All in all, we have been communicating the risks associated with the global atmospheric changes.

Many of the same difficulties we experience in alerting the public to the issues of atmospheric change arise whenever risk communication of hazardous substances is attempted; it is unclear who is actually at risk or what the risks are. The public has difficulty understanding the science and views the data as conflicting and confusing. Science does not have the data to answer some questions. Risk comparison data are sparse and poorly understood. There is much inaccurate and misleading information (1). ATSDR has developed a series of educational materials for health care professionals to give them a better understanding of the health risks of certain hazardous substances in the environment.

WHAT IS THE AGENCY FOR TOXIC SUBSTANCES AND DISEASE REGISTRY?

ATSDR is part of the Public Health Service of the United States Department of Health and Human Services. It was created by Congress in 1980 to implement the health-related sections of the Comprehensive Environmental Response, Compensation, and Liability Act (CERCLA or Superfund). The purpose of CERCLA is to protect the public from environmental exposures of hazardous substances. The mission of the Agency is to prevent or mitigate adverse human health effects and diminished quality of life resulting from exposure to hazardous substances in the environment. In 1986 the Superfund Amendments and Reauthorization Act (SARA) broadened the Agency's responsibilities in several areas including medical education. One of the goals of the Division of Health Education of ATSDR is to implement comprehensive and integrated educational programs for health providers on issues related to health care and exposure to hazardous substances.

HOW ARE RISKS COMMUNICATED?

The study of how risks are communicated is interesting. A phone survey conducted by the Georgetown University Medical Center (2) asked the public in six communities from across the country where it gets the bulk of its information about chemical and environmental risks. The results were revealing; they showed that risk and public health information is received primarily through the mass media (Table 1). For instance, the *Atlanta Journal* recently printed an article entitled "'80s had Earth's 6 Hottest Years." I am sure that your own local newspapers have printed similar stories.

The Georgetown study also asked whom we trust as information sources for hazardous substances in the environment. According to the results of the study, physicians are the most trusted, though not necessarily perceived as the most knowledgeable (Table 2). Such data indicate that, where environmental risks are concerned, we first trust the advice of physicians. But are physicians prepared to answer our questions about environmental health issues?

According to the recently published Institute of Medicine report *Role of the Primary Care Physician in Occupational and Environmental Medicine* (3), only a very small portion of the more than one-half-million physicians in this country are educated in environmental medicine. Environmental medicine is not included in the curriculum of most medical schools; there are no residency programs in environmental medicine in this country; and there is very little in the medical literature about environmental medicine (4). How can primary care physicians respond to patients' questions or diagnose environmental illnesses if they have little or no opportunity to receive information or training about it? Today's practicing physicians need to be provided with learning opportunities to bridge the gap between what they studied in medical school and what they need to know in order to respond to the concerns of their patients about hazardous substances in the environment. Similarly, they need to be informed that the environment may be a possible cause of the asthma, cataract, immune system dysfunction, skin cancer, and heatstroke cases that they treat, so that they are able to respond to patients' concerns about these illnesses.

CASE STUDIES IN ENVIRONMENTAL MEDICINE

Two subgoals of the Division of Health Education at ATSDR are to enhance the health professionals' recognition, treatment and prevention of illness or injury of persons exposed to hazardous substances, and to improve their ability to communicate health information concerning hazardous substances to their patients and the concerned public. Our division is developing a series of monographs to bridge the gap between the physicians' need-to-know and their opportunity to learn about environmental medicine. The series, *Case Studies in Environmental Medicine*, discusses hazardous substances on the U.S. Environmental Protection Agency's (EPA) priority pollutants list by presenting information based on what the practicing primary care physician needs to know.

Our Agency is preparing documents about the environmental issues of each of the priority pollutants. These *Toxicological Profiles* (5) discuss the chemical, biochemical, analytical, medical and public health information available about the hazardous substance. They can be obtained from the National Technical Information Service (Table 3). The medical information contained in the *Toxicological Profiles* is presented in condensed form in the *Case Studies in Environmental Medicine* series. To develop the *Case Studies*, we assembled a peer review committee with members from the American Medical Association, American Academy of Family Physicians, American College of Occupational Medicine, American College of Emergency Physicians, Association of State and Territorial Health Officers, and the Society of Teachers of Family Medicine. They, along with the contractor, DeLima Associates in San Rafael, California, developed the self-instructional format for the *Case Studies*.

Each *Case Study* opens with a current alert of significant, up-to-date information about environmental exposure to the hazardous substance, followed by the educational objectives for the issue. The practitioner is then presented with a composite case of a patient coming to the physician for diagnosis and treatment. The case report is written so that it can stand alone and be used in teaching or grand-rounds presentations. The hazardous substance involved in the case is not identified, but, instead, the reader is asked a series of pretest questions that were developed to promote further reflection upon the case. Didactic material is presented after the case report, organized by the following categories: exposure pathway, who is at risk, biological fate, physiological fate, symptoms and diagnosis, laboratory tests, treatment, standards and regulations, and resources. Throughout the didactic part of the monograph the reader is challenged with a series of questions designed to encourage application of the information to the case report. Answers to the pretest and challenge questions are provided at the end of the document. Also at the end is a post-test and evaluation that can be completed and returned for continuing education credit.

Reaction to the pilot evaluation has been very positive. Comments indicated that the documents are timely, well-written, well-targeted, and address the concerns of primary care physicians. Physicians participating in the pilot evaluation indicate that the documents are particularly useful because they do not require special equipment, such as computers or videotape players, and that they can be read at the physicians' leisure; time does not have to be taken away from their practice to attend a conference. We are currently investigating methods of dissemination including publication in selected medical journals.

While the intent of ATSDR is to develop documents concerning the hazardous substances found at hazardous waste sites, similar documents could be developed to alert physicians to the health effects of global atmospheric change. Before physicians can reliably diagnose and counsel patients about the health effects of atmospheric change, they must first be informed. The lack of adequate knowledge and skills to properly evaluate and advise patients with potential exposure to hazards can lead to serious problems with management of exposure situations. Underreaction can impede appropriate public response. Overreaction can impair the ability of the public to lead a normal, productive life, or accept qualified judgment regarding the significance of the hazard. We believe that the *Case Studies in Environmental Medicine* model is a mechanism that can effectively inform and educate physicians on these environmental issues.

Table 1. Source Of Recent Environmental Information

Source	Percent
Newspapers	76%
Local TV News	62
Radio	19
National TV News	11
Magazines	3
Government	2
Work	2
Family Members	2
Mail Notices	2

N = 669

Question: "In the past week, have you read or heard anything about the risks of chemicals or hazardous wastes in your area?" If yes, "Where did you read or hear this information?"

Source: McCallum, *et al.* (2)

Table 2. Perception of Information Sources

	Trust Percent Responding "A Lot"	Knowledge Percent Responding "Very"
News Reporters	27%	17%
Environmental Groups	40	53
Friends/Relatives	34	9
Local Emergency Planning Committee	28	33
State Government	12	29
Local Government	11	22
Federal Government	12	36
Chemical Industry Officials	8	58
Chemical Industry Employees	19	30
Physicians	46	27

Source: McCallum, *et al.* (2)

Table 3. Agency for Toxic Substances and Disease Registry
Toxicological Profile Information

Priority Group One Chemicals: Chemical (NTIS Order Number)

Aldrin/Dieldrin   Arsenic
Benzene   Benzo(a)anthracene
Benzo(a)pyrene   Benzo(b)fluoranthene
Beryllium   Cadmium
Chloroform   Chromium
Chrysene   Cyanide
Dibenzo(a,h)anthracene   1,4-Dichlorobenzene
Di(2-ethylhexyl)phthalate   Heptachlor/Heptachlor epoxide
Lead   Methylene chloride
Nickel   N-nitrosodiphenylamine
Polychlorinated biphenyls: 2,3,7,8-Tetrachlorodibenzo-p-dioxin
    Aroclor-1260, -1254, -1248,   Tetrachloroethylene
    -1242, -1232, -1221, and -1016   Trichloroethylene
Vinyl chloride

Priority Group Two Chemicals

Benzidine   Bis(2-chloroethyl)ether
Bis(chloromethyl)ether   Bromodichloromethane
Carbon tetrachloride   Chlordane
Chloroethane   p,p'-DDT, DDE, DDD
3,3-Dichlorobenzidine   1,2-Dichloroethane
1,1-Dichloroethene   1,2-Dichloropropane
2,4- & 2,6-Dinitrotoluene   Hexachlorocyclohexane
Isophorone   Mercury
N-nitrosodimethylamine   N-nitrosodi-n-propylamine
Pentachlorophenol   Phenol
Selenium   1,1,2,2-Tetrachloroethane
Toluene   1,1,2-Trichloroethane
Zinc

Each *Toxicological Profile* listed above was distributed in draft for public comment
in the last quarter of 1987 (Priority Group One) or during the first quarter of 1989
(Priority Group Two). These documents can no longer be obtained from the ATSDR in
draft form. The National Technical Information Service (NTIS) is responsible for the
distribution of finalized *Toxicological Profiles*, which become available as NTIS publi-

Table 3.   (continued)

cation numbers are assigned. Further information on the status of a *Toxicological Profile* listed above may be obtained by writing or calling:

> National Technical Information Service
> 5285 Port Royal Road
> Springfield, Virginia 22161
> (800) 336-4700 or (703) 487-4650

When all Priority Group One and Group Two *Toxicological Profiles* are finalized, a notice will be published in the *Federal Register* announcing their availability. The notice will include ordering instructions from the NTIS. There is a charge, determined by the NTIS, for these profiles.

Priority Group Three Chemicals

Acrolein	Acrylonitrile
Ammonia	Asbestos
Bromoform / Chlorodibromomethane	Chlorobenzene
Chloromethane	Copper
Creosote	Di-n-butylphthalate
1,2-Diphenylhydrazine	1,1-Dichloroethane
cis, trans 1,2-Dichloroethene	Endrin / Endrin aldehyde
Ethylbenzene	Ethylene oxide
Hexachlorobenzene	Naphthalene / 2-Methylnaphthalene
Nitrobenzene	Polycyclic aromatic hydrocarbons:
Plutonium	Acenaphthene, Acenaphthylene,
Radium	Anthracene,
Radon	Benzo(a)anthracene, Benzo(a)pyrene,
Silver	Benzo(b)fluoranthene,
Thorium	Benzo(g,h,i)perylene,
1,1,1-Trichloroethane	Benzo(k)fluoranthene, Chrysene,
2,4,6-Trichlorophenol	Dibenzo(a,h)anthracene
Uranium	Fluoranthene, Fluorene,
Total Xylenes	Indeno(1,2,3-cd)pyrene,
Toxaphene	Phenanthrene, Pyrene

The availability of each draft *Toxicological Profile* listed directly above was announced in the October 17, 1989 *Federal Register*. It is anticipated that these documents will be available from NTIS in December 1990.

Table 3.   (continued)

Priority Group Four Chemicals

Aluminum	Antimony
Barium	2,3-Benzofuran
Boron	Bromomethane
1,3-Butadiene	2-Butanone
Carbon disulfide	Cobalt
Cresols	Dibromochloropropane
1,2-Dibromoethane	2,4-Dichlorophenol
1,3-Dichloropropene	Endosulfan
Fluorides	2-Hexanone
Manganese	Methyl mercaptan
Methyl parathion	Mustard gas
Nitrophenol	Pyridine
Styrene	Thallium
Tin	1,2,3-Trichloropropane
Vanadium	Vinyl acetate

Each *Toxicological Profile* listed directly above is currently being developed by the ATSDR, and will be available as a draft for public comment in October 1990.

References

1. *Risk Communication and the Public Interest*. The Texas Risk Communication Project and The Woodlands Center for Growth Studies/Houston Area Research Center, October 1987.

2. McCallum, D.B., Hammond, S.L., Morris, L.A., Covello, V.T. *Public Knowledge and Perception of Chemical Risks in Six Communities: Analysis of a Baseline Survey*. Report prepared for U.S. Environmental Protection Agency, January 1990.

3. Institute of Medicine. *Role of the Primary Care Physician in Occupational and Environmental Medicine*. Washington, DC: National Academy Press, 1988.

4. Stoss, F.W. *Climate Change and Health – Information Trends and Resources*. Presented at Global Atmospheric Change and Public Health, December 5-6, 1989, Washington, D.C.

5. Agency for Toxic Substances and Disease Registry. *Toxicological Profile*. Series available from National Technical Information Service, Springfield, Virginia.

THE INFOTERRA NETWORK: A MODEL OF INTERNATIONAL INFORMATION EXCHANGE

Linda H. Spencer

INFOTERRA U.S. National Focal Point
U.S. Environmental Protection Agency
401 M Street, SW
Washington, D.C. 20460

Abstract

The international environmental referral system, known as INFOTERRA, was created by the United Nations Environment Programme following the United Nations Conference on the Human Environment held in Stockholm, Sweden, in June 1972. IN-FOTERRA has become the world's largest environmental information system with 136 National Focal Points (NFPs) covering 99% of the world's population. The NFPs link national and international institutions and experts in a cooperative venture to improve the quality of environmental decision making by providing timely global access to environmental data.

A founding member of the INFOTERRA network, the United States has remained one of the most active partners over the past decade. It is the second largest user of the INFOTERRA services and the number one provider of information. Recently, it has initiated a companion relationship with Botswana and the southern African region in an ongoing effort to strengthen network capabilities.

Introduction

In the fall of 1989, the U.S. Environmental Protection Agency (EPA) hosted Dr. Woyen Lee, Director of the United Nations Environment Programme's (UNEP) IN-FOTERRA system, at the national EPA Environmental Information Conference in Kansas City. Our purpose was to show him the information resources that could be accessed through the INFOTERRA U.S. National Focal Point (US NFP). After viewing all the impressive and highly technical exhibits of environmental information systems and databases, Dr. Lee simply shook his head. "It makes me very sad," he said, "as I realize how wide the gap in environmental information has grown between the developed and undeveloped world." That gap is a reality, and it is surely a fundamental concern to all of us involved in environmental information. The environmental health of this planet is

© 1990 by Elsevier Science Publishing Co., Inc.
Global Atmospheric Change and Public Health
James C. White, Editor

contingent upon both the global availability of accurate scientific and technological information and the ability to analyze that data and make sound environmental decisions. Both tasks require increasingly refined information technologies.

The subject of this conference – global atmospheric change and public health – provides reminders of both the interconnectedness of environmental issues and the problems those issues pose for the information community. For example, countries most affected by global atmospheric change, such as low-lying Bangladesh, are often those least equipped to access or analyze the information they need for adequate planning. In order to work together to address the environmental issues facing human civilization, we must begin with shared understandings and shared knowledge. When any country is hampered in environmental decision making efforts by a lack of access to information resources, all countries suffer.

The United Nations addressed this problem of access, equity and exchange of environmental information as early as 1972 at the World Conference on Human Environment in Stockholm, Sweden. When it created the United Nations Environment Programme, the Conference noted that information was a significant factor in environmental protection and mandated that an international environmental information system called INFOTERRA – "information of the earth" – be formed as the keystone of UNEP. INFOTERRA was charged with the task of providing an international organizational framework for sharing environmental information resources. In the seventeen years since 1972 that system has become an active productive network. (INFOTERRA, 1986)

Accessing the Network

INFOTERRA is now the world's largest environmental information system with 136 National Focal Points covering 99% of the world's population (see Annex 1 for list). The NFPs are responsible for the operation of the system within their country. They are the points of contact who receive queries from the user and who provide the INFOTERRA services. The NFPs link national and international institutions and experts in a decentralized cooperative environmental information exchange venture. Researchers who make inquiries to a National Focal Point are provided with information tailored to their needs including custom bibliographies, a selection of pertinent documents, lists of contacts, databases, video tapes, chemical samples or even such practical items as samples of drought resistant strains of seeds. In the quest for accurate and complete information, the NFP often contacts several sources. Additional resources within the INFOTERRA Network include *The INFOTERRA International Directory of Sources*, Special Sectoral Sources, and Regional Service Centres. These resources offer breadth to network activities. (INFOTERRA, 1989)

INFOTERRA's own *International Directory of Sources* is a rich compendium of 6,200 sources of environmental information from 91 countries. Registered by the National Focal Point, the sources have expressed a willingness to participate in international exchange and have demonstrated an ability to provide prompt, accurate and expert information in their subject area. Sixty percent of them are research institutes; twenty-eight percent are laboratories — with an equal percentage of institutions of higher learning; eighteen percent are libraries or documentation centers; and seventeen percent operate consultancy services. Collectively, they provide information on over 1,000 environmental subjects. (INFOTERRA, 1986) Since sources of information are by their nature dynamic and subject to change, each country updates its *Directory* listings every two years.

The *Directory* comes in four volumes including an index and is published biennially with a supplement issued in the intervening year. Sources are listed alphabetically by country and cross-referenced by subject. The *Directory* is available for purchase from the U.N. publications office. More than 45 NFPs search the *Directory* on microcomputer. The diskette version is available on UNESCO developed Micro-ISIS software.

To make the *Directory* more usable to the public, INFOTERRA has excerpted special interest areas from the larger collection and published smaller specialized directories. For example, there is a specialized directory on atmosphere and climate entitled *Man and the Atmosphere: Climate Change, Ozone Layer*. Other directories have been compiled in the areas of waste management, chemical safety, conservation for sustainable development, renewable sources of energy, and drinking water and sanitation.

Certain organizations and institutions are generally recognized as being leaders in their particular field of expertise. Twenty-five such organizations have agreed to serve as Special Sectoral Sources (SSS) and assist in the provision of substantive information to INFOTERRA users. The SSS draw upon their own expert knowledge, as well as any databases available to them, to provide custom-tailored, in-depth responses to queries on priority environmental problems. Requests for this information originate from the NFP and are channelled through the INFOTERRA Programme Activity Centre in Nairobi which maintains control over the volume and quality of the information flow and, subject to the availability of funds, pays any charges for users from developing countries. (Annual Report, 1988)

The SSS cover a wide variety of topics. The International Centre for Arid and Semi-Arid Land Studies at Texas Technological University and the Water and Sanitation for Health Project in Arlington, Virginia, are two U.S.-based examples (see Annex 2 for list). However, a number of subject areas are not represented and INFOTERRA is actively seeking support in several areas including both climate change and public health.

Countries within a geographic area share similar environmental concerns. The Regional Service Centre idea was conceived, first, to provide coordination of NFP activities within a geographic region and, second, to assist the lesser developed NFPs in their role as information providers. The nine Regional Service Centres (RSC) designated by the INFOTERRA are vehicles within each region for more efficient and economical use of information services. They coordinate computer searches of information sources, train focal point staff in the region, and devise and execute regional publicity campaigns (see Annex 3 for list). In an effort to deal with financial inequities among regions, INFOTERRA provides material or financial support to any needy RSC to enable it to assume an effective role. To date the system has been underutilized, and several subregions including our own are not linked to a RSC. Some, however, such as the one located in Chile, have filled a vital role linking information-resource poor nations to a strong information infrastructure.

The Network at Work

During 1988 over 15,000 inquirers from some 90 countries made use of the information services provided by INFOTERRA. (Drake, 1988) Over half of all queries were from users in developing countries. There are no typical questions or typical inquirers. The scope of the questions ranges from research methods on marine pollution to toxic chemical legislation, from the environmental effects of the use of bamboo in piping and plumbing to treatment methods for hospital waste, from fish breeding in industrial wastes to safe packaging materials for food – anything that falls under that all-encompassing term "environment." In recent years, however, growing concerns about pesticide residues and transboundary shipments of toxic waste have led to an increasing number of queries concerning the harmful effects of chemicals. Just as there are no typical questions, there are no typical questioners. School children, scientists, legislators, journalists, housewives and priests contact INFOTERRA. Whatever the inquiry and whoever the inquirer, the mandate of the NFPs is the same: provide substantive quality information in a timely fashion. Generally we are able to do just that.

Most of the inquirers – 83% according to the ongoing INFOTERRA user satisfaction survey – rated the services provided as excellent or very good. To illustrate with a few successes, the INFOTERRA Network has enabled the Republic of Gambia to develop hippo-proof fencing that saved large-scale rice paddy plantations from ruin, China to develop toxic chemical management regulations based on existing regulations from several countries, and Botswana to do an environmental impact assessment and reject the importation of large amounts of U.S. toxic waste for storage in the Kalahari Desert.

U.S. National Focal Point

A founding member of the INFOTERRA network, the United States has remained one of the most active partners over the past decade. It is the number one provider of information and the second largest user of the INFOTERRA services. The US NFP receives over 130 requests for information monthly. Sixty percent of these requests come from the developing world; of that subset, seventy-five percent deal with questions of chemical safety. Of the ten percent of inquiries originating in the U.S., the dominant questions concern environmental regulations that will affect business or investment opportunities abroad.

The US NFP is housed at the EPA Headquarters Library in Washington, D.C. This location enables the US NFP to work directly with EPA scientists and policy makers to ensure that answers to its requests are current and accurate. The focal point utilizes the extensive EPA library collection for its research activities. It also has direct access to the EPA library network, consisting of twenty-eight libraries located in the regional offices and laboratories across the country. These libraries provide a wide range of general and specialized information services. To assure that its research is comprehensive and current, the US NFP reaches beyond the confines of libraries to other sources through several hundred databases.

Future Development

INFOTERRA is seeking to strengthen the network by encouraging nations in the developed and developing world to form companion relationships for mutual assistance. The U.S. and Botswana NFPs have taken a lead in this effort by initiating such a relationship. Though still in the developmental phase, Botswana and the U.S. are jointly devising a strategy that will provide direct technological assistance needed by the Botswana NFP and aid the U.S. in developing a better understanding of Third World environmental realities and technology transfer issues. Together we envision a program that will extend beyond Botswana and enable the southern African region to work effectively as a sub-network within the INFOTERRA system.

Ireland and Uganda, Portugal and Portuguese-speaking Africa have followed our lead and are involved in developing companionships. Canada, Norway, Sweden and several countries in Latin America have also expressed an interest. The growing companion program should forge close and mutually-advantageous network ties. It should also transfer technology to the Third World which will result in a strengthened global environmental information exchange network that will be staffed by trained information specialists, and computerized and sensitized to the issue of appropriate technologies.

In the field of information and in the field of environment we need to share our experiences, our successes and failures, our research and its applications. In 1972 UNEP developed the operational tools for this cooperative endeavor, and in 1989 the INFOTERRA network is well established. INFOTERRA is an efficient global information system which enables both governments and nongovernmental bodies to utilize the knowledge and experience of others and, at the same time, to allow others to benefit from their own. It is an effective mechanism for the international exchange of environmental information. As we address the critical environmental problems of this decade, the INFOTERRA network will play an increasingly vital role. INFOTERRA invites you to take part. As a source of information or as a user of information you are most welcome to contact your National Focal Point.

References

Drake, S.F., 1988. *A Survey of U.S. Participation and Benefits*. Washington, D.C., Department of State: The United Nations Environment Programme.

United Nations, 1988. *Accis Guide to United Nations Information Sources on the Environment*. New York: United Nations.

United Nations Environment Programme, 1986. *INFOTERRA United Nations Environment Programme*. Nairobi, Kenya: United Nations Environment Programme.

United Nations Environment Programme, 1988. *Annual Report*. Nairobi, Kenya: United Nations Environment Programme.

United Nations Environment Programme, 1988, 1989. *INFOTERRA Bulletin*. Nairobi, Kenya: United Nations Environment Programme.

Annex 1. INFOTERRA Focal Points

012-Algeria
M.A. Mostefai
Point focal national INFOTERRA
Agence national pour la protection de l'environnement
49 Rue des Fusilles
El-Annesser
Alger
Algeria
Telephone: 77-14-14/77-14-39
Telex: 65439 ENL DZ

032-Argentina
Subsecretaria de Vivienda y Ordenamiento Ambiental
Avda Santa Fe 1548
10 Piso
Buenos Aires
Argentina
Telephone: 41-9721/42-5928

036-Australia
Mr. D. Macrae
INFOTERRA National Focal Point Manager
Department of Arts, Heritage and Environment
P.O. Box 1252
Canberra City ACT 2601
Australia
Telephone: (062) 467211
Telex: 62960 HOME AA

040-Austria
UNEP/INFOTERRA National Focal Point
Federal Ministry for Health and the Environment
Stubenring 1
Regierungsgebaude
A-1010 Vienna
Austria
Telephone: (0222) 57-56-55
Telex: 111780 REGEB

044-Bahamas
The Ministry of Health
P.O. Box N-3730
Nassau
Bahamas
Telex: 20264 EXTERNAL

048-Bahrain
Mr. Khalid Fakhro
Vice Chairman
Environment Protection Committee
P.O. Box 26909
Manama
Bahrain
Telephone: 275792
Telex: 8511 HEALTH BN

050-Bangladesh
Mr. M.A. Karim
Secretary
Environment Pollution Control Board
Lalmatia Housing Estate
House No. 6/11/F
Satmasjid Road
Dacca 7
Bangladesh
Telephone: 315777/318682

052-Barbados
The Permanent Secretary
Ministry of Tourism and the Environment
Culloden Farm
Culloden Road
St. Michael
Barbados
Telephone: 436-4820

056-Belgium
M.A. Stenmans
Président
Commission interministerielle de la politique scientifique
8 Rue de la Science
B-1040 Bruxelles
Belgium
Telephone: (2)230 41 00
Telex: 24501 PROSCIENT

084-Belize
Fisheries Administrator
Fisheries Department
Princess Margaret Drive
P.O. Box 148
Belize City
Belize

204-Benin
M.A. Acakpovi
Direction Generale de l'amenagement du territoire
    et de l'urbanisme
B.P. 239
Cotonou
Benin
Telex: 5200 MINAFF

068-Bolivia
Ing. Carlos Velasco Salguero
Director, SYFNID
Ministerio de Planeamiento y Coordinacion
Av. Arce. No. 2147
Casilla de Correo No. 8727
La Paz
Bolivia
Telephone: 372079
Telex: 3280 SYFNID
Cable: SYFNID La Paz

072-Botswana
Ms. B. Petto
INFOTERRA National Focal Point
NFP Manager
Private Bag 16
Francistown
Botswana

076-Brazil
Dr. Paulo Nogueira Neto
Secretario Especial do Meio Ambiente (SEMA)
Via W-3 Norte. Quadra 510. Bloco 08
Edificio Cidade de Cabo Frio
70.750 Brasilia D.F.
Brazil
Telephone: (61)274-9485
Telex: 611429 SEMA
Cable: SEMA BRASILIA

100-Bulgaria
Mr. I. Etropolski
Principal Director
Centre for Scientific and Technical Information
State Committee for Environment Protection
    Under Bulgarian Council of Ministers
Ul. Industrialna 7
1202 Sofia
Bulgaria
Telex: 22145 KOPS BGV

854-Burkina Faso
Mr. Zongo Joseph
Direction de l'aménagement forestier
    et du reboisement
Ministère des transports, de l'environnement
    et du tourisme
P.O. Box 7044
Ouagadougou
Burkina Faso
Telephone: 33-32-13
Telex: 5283 MINITOUR

108-Burundi
Mr. Kabayanda Audace
Directeur Général
Institut national pour la conservation de la nature
B.P. 2757
Bujumbura
Burundi
Telephone: 66-55

112-Byelorussian SSR
Byelorussian Scientific Research Institute of
    Economic Research of the State Plan (Belniinti)
BSSR State Planning Committee
Prospect Masherova 7
220676 Minsk
Byelorussian SSR
USSR

120-Cameroon United Republic
M. Njiensi Michel Ouakam
Secrétaire Permanent MAB et Point Focal INFOTERRA
Secrétariat Permanent Du Comité National MAB
B.P. 4742
Yaoundé
Cameroon United Republic
Telephone: 220562.223452

124-Canada
Ms. A. Bystram
Director, Departmental Branch
Department of Environment
Place Vincent Massey
Ottawa K1A 1C7
Canada
Telephone: (613) 997-2485
Telex: 533608 ENV HQ

132-Cape Verde
Mme. Betina Pais Santos
Centre de Documentation Technique et Scientifique
Case Postale 120
Praia
Cape Verde

140-Central African Republic
M.E. Mossona
Chef
Service de l'environnement
Ministère des travaux publics et de l'urbanisme
Bangui
Central African Republic

152-Chile
Attn: Jose Castella Arguelles
National Commission of Scientific and
    Technological Research (CONICYT)
Calle Canada No. 308
Casilla 297 V
Santiago
Chile
Telephone: 744537
Telex: 340191 CNCT CK

156-China, Peoples Republic
Mr. Wu Jin
Director, Institute of Environmental Chemistry
Chinese Academy of Sciences
P.O. Box 934
Beijing
China, Peoples Republic
Telephone: 28-5176/28-5129

170-Colombia
Hilda Dugand C., JEFE
Sección Información y Documentación
INDERENA
Diagonal 34 No. 5-18
Apartado Aereo No. 13458
Bogotá
Colombia
Telephone: 2340481.2831321
Telex: 44428 IND

174-Comoros
M. Abdou Moustakin
Direction de l'architecture et de l'urbanisme
Ministère de l'equipement et de l'environnement
B.P. 12
Moroni
Comoros

178-Congo
M. Issanga-Ngamissimi Marius
Directeur de l'environnement
Point focal national INFOTERRA
Direction de l'environnement
Ministère du tourisme, loisirs et de l'environnement
B.P. 958
Brazzaville
Congo
Telephone: 81-30-46

188-Costa Rica
Lic. Max Francisco Cerdas Lopez
Departamento de Información y Documentación
Consejo Nacional de Investigaciones Cientificas y Technológicas (CONICIT)
Apartado 10318
San José
Costa Rica
Telephone: 24-41-72
Cable: CONICIT

196-Cyprus
Nature Conservation Service
Ministry of Agriculture and Natural Resources
L. Akritas Ave.
Nicosia
Cyprus
Telephone: 40-2491
Telex: 4660 MINAGRI

200-Czechoslovakia
Ing. Ignac Fratric, CSC
Director
Czechoslovak Centre for the Environment – SVOP
Tr. Laca Novomeského 2
Bratislava 842 42
Czechoslovakia
Telephone: (07)326-462
Telex: 92229 WHOB C
Cable: UNOPOLCONT

208-Denmark
Mr. H. Sand
National Agency for Environmental Protection (Miljostyrelsen)
Strandgade 29
DK-1401 Copenhagen K
Denmark
Telephone: (01) 57-83-10
Telex: 31209 MILJOE DK

218-Ecuador
Ing. Carlos Velasco Salguero
Director Ejecutivo
Instituto Ecuatoriano de Obras Sanitarias
Toledo Sin Numero y Lerida
Casilla 680
Quito, Pinchincha
Ecuador
Telephone: 544-610

818-Egypt
Secretary
Council for Environmental Research
c/o Dr. Abou El-Fotoh Abdel-Lateif
Academy of Scientific Research and Technology
101 Kasr El Aine Street
Cairo
Egypt
Telephone: 31985
Telex: 93069 ASRT

222-El Salvador
Senor Ministro de Planificación
Secretario
Comité Nacional de Protección del Medio Ambiente
San Salvador
El Salvador
Telex: 30309 MIPLAN

230-Ethiopia
The National Revolutionary Development Campaign
    and Central Planning Supreme Council
The Physical Planning Department
P.O. Box 1037
Addis Ababa
Ethiopia
Telephone: 44-51-58/12-88-00

242-Fiji
The Director
Town and Country Planning Office
Government Buildings
P.O. Box 2350
Suva
Fiji
Telephone: 211759
Telex: 2167 FOSEC FJ
Cable: FOSEC FJ

246-Finland
Ms. Ann-Britt Ylinen
Ministry of the Environment
P.O. Box 306
SF-00531 Helsinki
Finland
Telex: 123275 SMPEL

250-France
M.R. Gimilio
Service de recherches études et du traitement de
    l'information sur l'environnement
Mission des systémes d'information
Ministère de l'environnement et du cadre de vie
14. Boulevard du Général Leclerc
F-92524 Neuilly Cedex
France
Telephone: 47.58.12.12
Telex: 620602 DENVIR

266-Gabon
M. Jean-Baptiste Mebiame
Directeur
Point Focal PNUE/INFOTERRA
Centre National Anti-Pollution
B.P. 3241
Libreville
Gabon
Telephone: 73-17-07. 73-08-73

270-Gambia
Mr. Serigne Omar Fye
Environmental Officer
Ministry of Water Resources and Environment
5 Marina Parade
Banjul
Gambia
Telephone: 431. 537
Telex: 204 PRESOR GV

278-German Democratic Republic
Director
Centre for Protection and Improvement of Environment
Ministry for Environment Protection and Water Management
Schnellerstrasse 140
DDR-1190 Berlin
German Democratic Republic
Telephone: 638920
Telex: 112368 1FW DD

280-Germany, Federal Republic
Ms. Ines Schusdziarra
Ministerialrat
Bundesministerium des Innern
Referat UI 4
Graurheindorferstrasse 198
D-53 Bonn 7
Germany, Federal Republic
Telephone: (228)6811-4151

288-Ghana
Mr. Gyamfi-Aidoo
Programme Officer
Environmental Protection Council
P.O. Box M326
Accra
Ghana
Cable: ENVIRON ACCRA

300-Greece
Ms. Pinelopi Nicolaou
Electronic Engineer
Scientific Collaborator in the Ministry of Physical Planning,
    Housing and Environment
Amaliados Street 17
Ambelokipi
Athens
Greece
Telephone: 644-4143
Telex: 216374 IHOP

320-Guatemala
Coordinador de la Comision Asesora del Presidente
Comisión Ministerial Encargada de la Conservacion
    y Mejoramiento del Medio Humano
Oficina 10, 1er Piso
Ed. Registro de la Propiedad
9A Avda Entre 14 y 15 Calles
Zona 1
Cuidad de Guatemala
Guatemala
Cable: MINISGOB GUATEMALA

324-Guinea
Mme. Aissatou Bah
Point focal national INFOTERRA
Direction de la recherche scientifique et technique
Secretariat d'état à l'enseignement superieur
    et de la recherche scientifique
B.P. 561
Conakry
Guinea
Telephone: 46-10-12/46-19-71

624-Guinea-Bissau
Ms. A-M de sa Almeida
Directrice des Forêts et Chasse
Ministère des Ressources Naturelles
CP 399
Bissau
Guinea-Bissau

328-Guyana
Ms. M. Taylor
Officer — INFOTERRA
Environmental Unit
Inst. of Applied Science and Technology
44 Père Str. Kitty
Greater Georgetown
Guyana
Telephone: 53-822/53-829/62-15

340-Honduras
Lic. Wilberto Aguilar N.
Jefe
Departamento de Vida Silvestre
Dirección General de Recursos
Naturales Renovables
Boulevard Toncontin
Barrio Guacerique
Comayaguela D.C.
Honduras
Telephone: 22-5782

344-Hong Kong
Secretary for Health and Welfare
Health and Welfare Branch
7th Floor, Government Secretariat
Lower Albert Road
Hong Kong
Telephone: 5-95391
Telex: 73380 GOVHK

348-Hungary
National Authority for Environment Protection
    and Nature Conservation
P.O. Box 732
H-1531 Budapest
Hungary
Telephone: 327-739
Telex: 227607/8 OKTH H
Cable: OKTH BUDAPEST

356-India
Mr. Harjit Singh
Director (Envis)
Department of the Environment
Bikaner House
Shahjahan Road
New Delhi 110011
India
Telephone: 306156
Telex: 315265 DOE
Cable: PARYAVARAN NEW DELH

360-Indonesia
National Scientific Documentation Centre (PDIN)
Jl. Gatot Subroto
P.O. Box 3065
Jakarta Selatan
Indonesia
Telephone: 583465/66/67
Cable: PDIN

368-Iraq
Dr. I.M. Al-Samawi
Director General
Human Environment
Oquba Bin Nafi Square
Baghdad
Iraq
Telephone: 98827
Telex: 2707 HEALTH IK
Cable: BEAM

372-Ireland
Mr. Noël Hughes
Library Officer
National Institute for Physical Planning
    and Construction Research
St. Martin's House
Waterloo Road
Dublin 4
Ireland
Telephone: (01)602511
Telex: 30846 FORB EI

376-Israel
Ms. Debra Savitzky
Director-Israel INFOTERRA NFP
Environmental Protection Service
Ministry of the Interior
P.O.B. 6158
Jerusalem 91061
Israel
Telephone: (02) 660151 Ext. 285
Telex: 26162 IEPS IL
Cable: MEM PNIM EPS

380-Italy
Mr. A. Candeloro
Direzione Generale Servizi Igiene Pubblica
Ministero Della Sanita
Via Liszt 34
1-00100 Rome (EUR)
Italy
Telephone: (06) 591-6941
Telex: 610453 MINISAN

384-Ivory Coast
Mlle. Therese Lokpo
Commission nationale de l'environnement
B.P. V-67
Abidjan
Ivory Coast
Telephone: 22-53-54
Telex: 23399 MINIMAR
Cable: MINIMA

388-Jamaica
INFOTERRA National Focal Point
Natural Resources Conservation Department (NRCD)
Ministry of Agriculture
53 1/2 Molynes Road
P.O. Box 305
Kingston
Jamaica
Telephone: 923-5155/923-5070

392-Japan
Shota Hirosaki
Director
Environmental Information Division
The National Institute for Environmental Studies
16-2 Onogawa. Yatabe-Machi
Tsukuba-Ibaraki 305
Japan
Telephone: 0298-51-6111
Cable: KOGAIKEN TSUKUBA

400-Jordan
H.E. The Minister for Planning
Ministry for Planning
P.O. Box 555
Amman
Jordan
Telex: 21319 NPC

404-Kenya
Director
National Environment Secretariat
Ministry of Environment and Natural Resources
Kenyatta International Conference Centre
P.O. Box 67839
Nairobi
Kenya
Telephone: 332383 Ext. 2102

410-Korea, Republic
Mr. Yun-Syng, Cho
Officer in Charge - INFOTERRA
Chief, Environmental Health Division
National Environmental Protection Institute — NEPI
280-17 Pulgwang-Dong Unp' yong-qu
Seoul
Korea, Republic
Telephone: 385-5711/20
Telex: 25783 ENVIROK

414-Kuwait
Ms. Mariam Al Awadi
Director
Ministry of Planning
P.O. Box 15
Kuwait
Telex: 23433 CONF

422-Lebanon
Mr. Malek Salam
Council for Development and Reconstruction
Presidence
Baabda
Lebanon
Telex: 21000 PRL

430-Liberia
Mr. T. Teage Jr.
Assistant Minister for Research and Development Planning
Ministry of Local Government
Monrovia
Liberia
Telex: 4374 MINIPLAN

434-Libyan Arab Jamahiriya
Mr. Farhat-Shawashi
Director-General of Technical Department
Secretariat of Utilities
Tripoli
Libyan Arab Jamahiriya
Telephone: 35838
Telex: 20122 BALADY

450-Madagascar
M.D. Ratsimbazafy
Chef. Salle Ops
Ministère du developpement rural et reforme agraire
B.P. 301
Antananarivo
Madagascar
Telex: 22508 MPARA

454-Malawi
The Secretary
National Research Council
Office of the President and Cabinet
Private Bag 301
Lilongwe 3
Malawi
Telephone: 731199
Telex: 4389 PRES MI
Cable: PRESMIN LILONGWE

456-Malaysia
The Director General
INFOTERRA NFP
Department of Environment
Ministry of Science, Technology and Environment
13th Floor, Wisma Sime Darby
Jalan Raja Laut
40662 Kuala Lumpur
Malaysia
Telephone: 603-293886
Telex: 28154 MOSTEC MA

466-Mali
Mr. Salif Kanouté
Directeur Général Adjoint des Eaux et Forêts
Ministère de l'Elevage et des Eaux et Forêts
Bamako
Mali
Telephone: 225850/225973

470-Malta
INFOTERRA National Focal Point
Malta Human Environment Council
Ministry of Health
15 Merchants Street
Valletta
Malta
Telephone: 24071
Telex: 1100 MODMLT MT
Cable: HEALTH MALTA

478-Mauritania
Chef du Service de la Protection de la Nature
B.P. 170
Nouakchott
Mauritania
Telex: 585 MTN

480-Mauritius
Mr. T. Ramyead
Ministry of Housing, Lands and Environment
Edith Cavell Street
Port Louis
Mauritius

484-Mexico
Centro de Coordinación Nacional INFOTERRA
Subsecretaria de Ecologia
Secretaria de Desarrollo Urbano y Ecologia
Cuidad de México
México
Telephone: 553-96-89

496-Mongolia
Tsentr Nauchnoi 1 Tekhnicheskoi Informatsii
Ul. Kolarova 49
Ulan Bator
Mongolia
Telephone: 20334
Telex: 245 UL-BT

504-Morocco
M.A. Fassi-Fihri
Point focal INFOTERRA
Centre national de documentation
Charii Maa Al Ainain
Haut Agdal
B.P. 826
Rabat
Morocco
Telephone: 749-44
Telex: 31052 CND M

524-Nepal
National Committee for Man and the Biosphere
c/o Ministry of Education and Culture
Kaisher Mahal
G.P.O. Box 1071
Kathmandu
Nepal
Cable: NEPNATCOM

528-Netherlands
Mr. Laurens de Lavieter
Study and Information Centre on Environmental
    Research (TNO)
TNO Complex Zuidpolder
Postbus 186
NL-2600 Delft
Netherlands
Telephone: (015) 56-93-30
Telex: 38071 ZPTNO NL

554-New Zealand
Mr. K. Piddington
Commission for the Environment
CPD House
108 The Terrace
P.O. Box 10241
Wellington N-2
New Zealand
Telephone: 720642
Telex: 3276 RESEARCH

562-Niger
Direction hygiène et medecine mobile
B.P. 371
Niamey
Niger

566-Nigeria
Dr. Raimi O. Ojikutu
Director
Environmental Planning and Protection Division
Federal Ministry of Works and Housing
4 Shaw Road
Ikoyi
Lagos
Nigeria
Telephone: 682625
Cable: PERMHOUSE LAGOS

578-Norway
International Division
Ministry of the Environment
Myntgt 2 – Dep.
Oslo 1
Norway
Telephone: (2)41-90-10 Ex. 7524
Telex: 18990 ENV N

512-Oman
Council for Conservation of Environment
    and Water Resources
P.O. Box 5310
Ruwi
Oman
Telephone: 704344/704346
Telex: 3590 ENVRMNT ON

586-Pakistan
Dr. Aejaz Ahmed Malik
Project Director
Pakistan Scientific and Technological Information
    Centre (Pastic)
Quaid-I-Azam University Campus
P.O. Box 1217
Islamabad
Pakistan
Telephone: 824161, 811375
Cable: PASTIC ISLAMABAD

590-Panama
Centro de Coordinación Pnuma/INFOTERRA
Ministerio de Planificación y Política Económica
Via Espana, Edificio Ogawa
Panamá 4
Panamá
Telex: 2738 MIPPE

598-Papua New Guinea
Ms. M. Mark
Department of Environment and Conservation
Central Government Offices, Waigani
P.O. Box 6601
Boroko
Papua New Guinea
Telex: 22144 NE

600-Paraguay
National Service for Environmental
    Sanitation (SENASA)
Asunción
Paraguay

604-Peru
Sr. Director General Nacional
Oficina Nacional de Evaluación de Recursos
    Naturales (ONERN)
Calle Diecisiete No. 355
Urb. El Palomar
San Isidro
Apartado 4992
Lima
Peru
Telephone: 414606/457689
Cable: ONERN LIMA-PERU

608-Philippines
Program Manager
Technobank
Technology Resource Centre
TRC Building
Buendia Avenue Ext.
Makati
Metro Manila
Philippines
Telephone: 88-98-11
Telex: 64002 TRC PN
Cable: TECHCENTER MANILA

616-Poland
Dr. Hanka Zanieweska
Director
Instytut Ksztaltowania Srodowiska
Ul. Krzywickiego 9
02-078 Warsaw
Poland
Telephone: 21-64-81
Telex: 813493 IKS PL

620-Portugal
Sr. J.C.M. da Cunha
President
Comissao Nacional do Ambiente
Praca Duque de Saldanha 31-5
1096 Lisboa Codex
Portugal
Telephone: 54-40-25
Telex: 18462 CNAMBI P
Cable: CNAMBI LISBOA

634-Qatar
Mr. Abdul Rahman Mohammad Bin Jabor Al-Thani
Deputy Chairman
Environment Protection Committee
P.O. Box 7634
Doha
Qatar
Telephone: 32 08 25
Telex: 4579 EPC DH

642-Romania
Dr. V. Ianovici
President
Conseil national pour la protection
    de l'environnement
Boulevard Ilie Pintilie Nr. 5
Bucarest
Romania

646-Rwanda
M.J. Zigirababili
Point focal INFOTERRA/Rwanda
c/o Ministère de l'industrie, des mines
    et de l'artinasat
BP. 73
Kigali
Rwanda
Telephone: 5916

882-Samoa
Mr. Iosefatu Reti
Chief Forestry Officer
Forestry Division
Department of Agriculture, Forests and Fisheries
P.O. Box 206 Apia
Samoa
Telephone: 22 561 - 49
Telex: 233 TREASURY SX

682-Saudi Arabia
Director General
General Directorate of Meteorology
Ministry of Defence and Aviation
P.O. Box 1358
Jeddah
Saudi Arabia
Telephone: 6654188/6692288
Telex: 401236 ARSAD SJ
Cable: ARSADEPT JEDDAH

686-Senegal
M.O. Diop
Directeur du Centre national de documentation
    scientifique et technique
Secretariat d'État à la Recherche Scientifique
    et Technique
61 bvd. Pinet-Laprade
B.P. 3218
Dakar
Senegal
Telephone: 21-51-63/22-44-75

690-Seychelles
Mr. A.J.P. Roux
Secretary
Seychelles National Environment Commission
Ministry of National Development
Independence House
P.O. Box 199
Victoria
Mahé
Seychelles
Telephone: 22881
Telex: 2312 MINDEV
Cable: MINDEV SEYCHELLES

706-Somalia
Mr. Mohamed Salaad Samatar
INFOTERRA National Focal Point
Head, Multilateral Co-operation Service
Directorate of Economic Co-operation
Ministry of Foreign Affairs
Mogadishu
Somalia

724-Spain
Sr. Manuel Matesanz Santamaria
Coordinador Nacional de INFOTERRA
Direccion General de Medio Ambiente
Ministerio de Obras Publicas y Urbanismo
Paseo Castellana 67
Madrid 3
Spain
Telephone: 254-7630
Telex: 22325 MINOP

144-Sri Lanka
Mr. V.K. Nanayakkara
Director (Secretariat)
Central Environmental Authority
Maligawatte New Town
Colombo 10
Sri Lanka
Telephone: 549455/549456

662-St. Lucia
Deputy Director
Ministry of Finance and Planning
Government Buildings
P.O. Box 709
St. Lucia
Telephone: 23688
Telex: 6243 PMSLU LC

736-Sudan
Director
National Documentation Centre
National Council for Research
General Secretariat
P.O. Box 2404
Khartoum
Sudan
Telephone: 70776/70702
Telex: 22342 ILMI
Cable: BUHUTH

752-Sweden
INFOTERRA/Sweden
Dr. Arne Sjöqvist
National Environmental Protection Board
Box 1302
S-171 25 Solna
Sweden
Telephone: (08) 799-10-00
Telex: 11131 ENVIRON
Cable: ENVIRON S

756-Switzerland
Office Fédéral de la protection de l'environnement
Service des organisations internationales
CH-3003 Berne
Switzerland
Telephone: (031) 61-93-23
Telex: 911191 HELV CH
Cable: HELV CH

760-Syrian Arab Republic
Ministry of State for Environmental Affairs
Government House
Damascus
Syrian Arab Republic
Telephone: STD 226-000/210-011
Telex: 411930 SYTROL

834-Tanzania, United Republic
Mr. G.L. Kamukala
Head
Environment Protection and Management Section
Town Planning Division
Ministry of Lands, Natural Resources and Tourism
P.O. Box 20671
Dar es Salaam
Tanzania, United Republic
Telex: 41284

764-Thailand
Dr. Pravit Ruyaphorn
Secretary-General
National Environment Board
Soi Prachasumpun 4
Rama VI Road
Bangkok 4
Thailand
Telephone: 278-5467
Telex: 20838 MOSTE
Cable: NEB BANGKOK

768-Togo
M. Anani Agbekodo
Division Environnement
Direction forêts, chasse et environnement
Ministère de l'aménagement rural
B.P. 355
Lomé
Togo

788-Tunisia
Mme. Hedia Baccar
Ministère de l'agriculture et de l'environnement
30 Rue Alain Savary
Tunis
Tunisia
Telephone: 890863/890926
Telex: 13378 MINAGR TN

800-Uganda
Mr. Michael Werikhe
Ministry of Housing and Regional Planning
P.O. Box 1911
Kampala
Uganda
Telephone: 43543 34350
Cable: PLANNING KAMPALA

804-Ukrainian SSR
INFOTERRA National Focal Point
Ukrainian Research Institute of Scientific and
      Technical Economic Studies of Ukrainian
      State Planning Committee
Kiev
Ukrainian SSR
USSR

784-United Arab Emirates
Dr. Abdul Wahab Muhaideb
Deputy Chairman
The Higher Environment Committee
Ministry of Health
P.O. Box 848
Abu Dhabi
United Arab Emirates
Telephone: 324742
Telex: 22678 MEDAB EM
Cable: MEDAB ABU DHABI

826-United Kingdom
Mrs. J.A. Deschamps
UK/INFOTERRA – Room P3/008D
Department of the Environment
2 Marsham Street
London SWIP 3EB
United Kingdom
Telephone: (01) 212-5270
Telex: 22221 DOEMAR

840-United States
U.S. NFP INFOTERRA (PM 211A)
U.S. Environmental Protection Agency
401 M Street SW
Washington, DC 20460
United States
Telephone: (202) 382-5917
Telex: 892758 EPA WSH
Cable: EPAWSH

858-Uruguay
Instituto Nacional para la Preservación
    del Medio Ambiente
Secretaria de Comisiones
Sarandi 444 Piso 2
Montevideo
Uruguay
Telephone: 950103 INT 39

810-USSR
Mr. A.I. Mikhailov
VINITI
14 Baltijskaya Ul.
125219 Moscow
USSR
Telephone: 1554250
Telex: 411249 VINITI

862-Venezuela
Lic. Ines Tabares de d'Ambrosio
Directora de Informatica
Coordinadora Nacional de INFOTERRA
Ministerio del Ambiente y los Recursos
    Naturales Renovables
Torre sur Piso 11
Centro Simón Bolívar
Caracas 1010
Venezuela
Telephone: (02)41-42-10
Telex: 21378 MARNR VEN
Cable: MARNR

866-Viet Nam
Mr. Nguyen Van Khanh
Comité d'État des sciences et techniques
Centre INFOTERRA du Viet Nam
39 rue Tran Hung Dao
Hanoi
Viet Nam
Telephone: 52731
Telex: 287 UBKHT VNN

886-Yemen
Ministry of Municipalities
Environment Department
P.O. Box 1445
Sana'a
Yemen
Telephone: 215658
Telex: 2526 MUHAWS YE

890-Yugoslavia
Conseil de l'Environnement et de l'Amenagement du Territoire
Savezno Izvrsno Vece
Bulevar Lenjina 2
11070 Belgrade
Yugoslavia
Telephone: 638-253

180-Zaire
Point focal INFOTERRA/Zaïre
Département de l'environnement, conservation de la nature et tourisme
B.P. 12.348
Kinshasa 1
Zaire

894-Zambia
Documentation and Scientific Information Centre
National Council for Scientific Research
P.O. Box Ch. 158
Chelston
Lusaka
Zambia
Telephone: 75321/2/3/4/5/6
Cable: NACSIR CHELSTON ZAM

716-Zimbabwe
The Permanent Secretary
Ministry of Natural Resources and Tourism
5th Floor Mukwati Building
Private Bag 7753, Causeway
Harare
Zimbabwe
Telephone: 705671 Ext. 27
Telex: ZW 4435 ZIMTO

902-International
The Director
INFOTERRA Programme Activity Centre
United Nations Environment Programme
Box 30552
Nairobi
Kenya
Telephone: 33-39-30
Telex: 22173 UNEP
Cable: UNITERRA NAIROBI

Annex 2.   List of INFOTERRA Special Sectoral Sources (SSS)

INFOTERRA Special Sectoral Sources (SSS)	Subject Covered
1. Mr. J. Huismans Director International Register of Potentially Toxic Chemicals (IRPTC) United Nations Environment Programme (UNEP) Palais des Nations 1211 Geneva 10 Switzerland Telex: 28877 UNEP CH Telefax: 3398 77 and 33 98 79 (Message should begin "FOR IRPTC"...")	Toxic chemicals
2. Industry and Environment Office (IEO) United Nations Environment Programme (UNEP) Tour Mirabeau 39/43 quai André Citroën 75739 Paris Cedex 15 Telex: 204997 Telephone: 45.78.33.33 Telefax: 45 78 32 34 (COFRAMINES) FOR UNEP	Industry and environment
3. Ms. F. Burhenne-Guilmin Head Environmental Law Information System (ELIS) International Union for Conservation of Nature and Natural Resources (IUCN) Adenauralle 214 D-53 Bonn 1 Federal Republic of Germany Telex: 886808W BTAG D	Environmental legislation
4. Director Environmental Sanitation Information Centre (ENSIC) Asian Institute of Technology (AIT) P.O. Box 2754 Bangkok Thailand 10501 Telex: 84276 AIT TH Telefax: 66 2 529 0374	Water supply and sanitation

5. Director   New and
    Renewable Energy Resources   renewable
        Information Centre (RERIC)   energy
    Asian Institute of Technology (AIT)
    P.O. Box 2754
    Bangkok
    Thailand 10501
    Telex: 84276 AIT TH
    Telefax: 66 2 529 0374

6. Professor H. Dregne   Arid lands and
    International Centre for Arid and   desertification
        Semi-arid Land Studies (ICASALS)
    Texas Technological University
    Box 4620
    Lubbock, Texas 79409
    USA
    Telex: 910 896 4398 TTU CID LBK
    Telefax: 0001-806-742-1900

7. Mr. S.R. Samady   Environmental
    Director   education and
    Division of Science, Technical and   training
        Vocational Education
    UNESCO
    7, place de Fontenoy
    75700 Paris
    France
    Telex: 204661

8. Mr. A.S. Bhalla   Appropriate
    Chief   technology
    Technology and Employment Branch
    Employment and Development Department
    International Labour Office (ILO)
    Geneva
    Switzerland
    Telex: 22271 BITGE
    Telefax: 4122-98 86 85

9. Mr. H.W. Pack   Appropriate
    Chief   technology;
    INTIB Unit   industry and
    Industrial Information Section   environment
    UNIDO
    Vienna International Centre
    P.O. Box 300
    A-1400 Vienna
    Austria
    Telex: 135612
    Telefax: 232156

10. All-Union Scientific and Research   Hydrometeo-
        Institute of Hydrometeorological   rology
        Information
    International Data Centre VNIIGMI-MTsD
    USSR State Committee of Hydrometeorology and
        Control of Natural Environment
    249020 Kaluga District
    Obninsk
    Koroleva St. 6
    USSR
    Telefax: 255-66-84

11. INFOTERRA SSS   Environmental
    Centre for Environmental Management   management and
        and Planning (CEMP)   planning (EIA)
    Department of Geography
    University of Aberdeen
    High Street
    Old Aberdeen AB9 2UF
    Scotland
    United Kingdom
    Telex: 73458 UNIABN G
    Telefax: 0224 491439

12. Ms. M. Lund — Waste management
    Commercial Office
    Marketing and Sales Department
    United Kingdom Atomic Energy Authority
    Harwell Laboratory
    Bldg. 7.12
    AERE Harwell
    Oxfordshire, OX11 ORA
    United Kingdom
    Telex: 83135 ATOMHA G
    Telefax: (0235) 832591 and (0235) 432916

13. Global Resource Information Database (GRID) — Environmental monitoring and assessment
    Global Environmental Monitoring System (GEMS)
    United Nations Environment Programme (UNEP)
    P.O. Box 30552
    Nairobi
    Kenya
    Telex: 22068 UNEP KE

14. Dr. D. Mentz — Agriculture in relation to environment (food-control of pests and diseases, post-harvest loss, etc.)
    Director-General
    C.A.B. International
    P.O. Box No. 100, Wallingford
    Oxon OX10 8DF
    United Kingdom
    Telephone: (0421) 32111
    Telex: 847964 (COMAGG G)
    Telefax: (0491) 33508

15. Mr. N. Hughes — Transportation and environment
    Library Officer
    National Institute for Physical Planning and
        Construction Research
    St. Martin's House
    Waterloo Road
    Dublin 4
    Ireland
    Telex: 30846 FORB EI

16. Mr. Jeffrey A. McNeely   Conservation
    Director   and environment
    Programme and Policy Division
    International Union for Conservation of Nature and
        Natural Resources
    Avenue du Mont-Blanc
    CH-1196 Gland
    Switzerland
    Telex: 22618 IUCN CH

17. Mr. J.L. MacLean   Living aquatic
    Director   resources
    Information Program   management
    The International Center for Living Aquatic Resources
        Management
    3rd Floor
    Bloomingdale Building
    Salcedo Street
    P.O. Box 1501
    Makati
    Metro Manila
    Philippines
    Telex: ITT 45658 ICLARM PM
    Telefax: (63-2) 819-3329 Attn. ICS 406

18. The Australian Institute of Marine Science,   Marine
        with the Great Barrier Reef Marine   environment
        Park Authority and James Cook University
        of North Queensland
    PMB No. 3
    Townsville MC Queensland 4810
    Australia
    Telex: 47009 UNITOWN — (re. query — "Attn.
        Ms. I. Bush, Librarian)
    Telefax: 61 077 725852

19. Dr. R.K. Pachauri   Energy and
    Director   environment
    Tata Energy Research Institute
    7, Jor Bagh
    New Delhi 110003
    India
    Telex: 31-61593 TERI IN

20. Mr. Leo A. St. Michel   Water supply
    Project Director   and sanitation
    Water and Sanitation for Health Project (WASH)
    WASH Operations Center
    1611 N. Kent Street
    Room 1002
    Arlington
    Virginia 22209
    USA
    Telex: WUI 64552

21. Mr. Marcel Marchand   Environment and
    EDWIN Project Leader   development
    Centre for Environmental Studies
    Garenmarkt 1b
    P.O. Box 9518
    2300 RA Leiden
    The Netherlands
    Telex: (UNIV.): 39427 BURUL
    Telephone: 31/71-277470/277486

22. Mr. Henrique B. Cavalcanti   Industry and
    Director   environment
    International Environmental Bureau
        (A Specialized Division of the ICC)
    61, Route de Chêne
    Geneva CH 1208
    Switzerland
    Telex: 289 556
    Telephone: (22) 865111
    Telefax: (22) 36-03-36 (via LAURENS)

23. Mr. J. Takala   Occupational
    Head   safety and
    International Occupational Safety and   health
        Health Information Centre (CIS)
    International Labour Office (ILO)
    CH 1211 Geneva 22
    Switzerland
    Telex: 22.271 BIT CH
    Telephone: (22) 99 67 40

24. Mr. S. Schrader   Afforestation
    Head   and
    Documentation Centre   reforestation
    Bundesforschungsanstalt für Forst — und
        Holzwirtschaft
    Leuschnerstr. 91, Postfach 80 02 10
    2050 Hamburg 80
    Federal Republic of Germany
    Telephone: (040) 739 62 x 435
    Telex: 403826 – BFH

25. Director   Agricultural
    All-Union Research   microbiology
    Institute for Agricultural Microbiology
    Podbelsky Shosse, 3
    188620 Leningrad-Pushkin
    USSR

Annex 3.   List of INFOTERRA Regional Service Centres

1. Ms. Pippa Smith   For Southeast
    INFOTERRA National Focal Point   Asia and South
    Department of the Arts, Sport, the   Pacific Sub-
        Environment, Tourism and Territories   regions
    GPO Box 787
    Canberra City, ACT 2601
    Australia
    Telex: 61716

2. Ms. Ana Maria Pratt Trabal   For Latin
    Director of Information   America and
    National Commission of Scientific and   the Caribbean
        Technological Research (CONICYT)   Region
    Calle Canada No. 308
    Casilla 297 V
    Santiago 21
    Chile
    Telex: 340191 CNCT CK

3. Mr. Hargit Singh   For Southern
    Director (ENVIS)   Asia Sub-region
    Department of Environment and Forests
    B Block, Paryavaran Bhavan
    CGO Complex, Lodi Road
    New Delhi – 110003
    India
    Telex: W-66185 DOE IN

4. Mr. A. Fassi-Fihri   For Northern
    Director   Africa and
    Point focal INFOTERRA   Western Asia
    Centre national de Documentation   Sub-regions
    Charii Maa Al Ainain
    Haut Agdal
    B.P. 826
    Rabat
    Morocco
    Telex: 31052 CND M

5. Ms. Faria El-Zahawi   For Arab League
    Director   Countries
    Documentation and Information Center
    League of Arab States
    37, Bd. Kheir Eddine Bacha
    Tunis
    Tunisia
    Telex: 14 412 JAMIA TN

6. Mr. A.V. Butrimenko   For CMEA
    Director   Countries
    International Centre for Scientific and   and Yugoslavia
        Technical Information (ICSTI)
    21-B Kuusinen Street
    125252 Moscow
    USSR
    Telex: 411925 ICSTI, MOSCOW SU

7. Director   For Eastern,
    National Environment Secretariat   Central and
    Ministry of Environment and Natural Resources   Southern
    Kenyatta International Conference Centre   African Sub-
    P.O. Box 67839   regions
    Nairobi
    Kenya

8. M.O. Diop   For French-
    Directeur du Centre national de documentation   speaking
        scientifique et technique (CNDST)   Central and
    Ministère du plan et de la coopération   Western
    61 Boulevard Pinet – Laprade   African
    B.P. 3218   Countries
    Dakar
    Senegal
    Telex: 3133 PLAN-COOP

9. Lic. Ines Tabares de d'Ambrosio   For the
    Directora de Informática   Caribbean
    Coordinadora Nacional de INFOTERRA   Sub-region
    Ministerio del Ambiente y de los Recursos
        Naturales Renovables
    Torre sur – Piso 11
    Centro Simón Bolívar
    Caracas 1010
    Venezuela
    Telex: 21378 MARNR VEN

INFORMATION SOURCES AND NEEDS

Geoff Worton

American Institute of Aeronautics and Astronautics
Technical Information Service
555 West 57th Street, Suite 1200
New York, NY 10019

Current changes occurring in the Earth's atmosphere can be expected to produce serious health hazards to the world's population. These hazards may range from malnutrition as a result of prolonged regional famine and drought to skin cancer and eye disease caused by exposure to increased levels of ultraviolet light.

In order to deal with these potential problems, scientists and researchers will require more and different kinds of constantly updated scientific and technical information. And they will require that information to be available quickly, accurately, comprehensively and manageably. This paper addresses some of the information sources available from the American Institute of Aeronautics and Astronautics' Technical Information Service and how those sources can benefit research.

The American Institute of Aeronautics and Astronautics (AIAA) is a 45,000 member professional society for engineers and scientists working in aerospace.

AIAA's information resources include:

- Technical meetings and conferences
- Primary publications and monographs
- Educational programs
- Technical information service

Services from AIAA's Technical Information Service include the Aerospace Database, an online resource, and the AIAA Library.

The Aerospace Database contains abstracts and full bibliographic citations for world-wide literature relating to aerospace in the broadest definition of the term (for a list of the subjects covered, please refer to the appendix). Literature sources include:

- Journals
- Conference papers and proceedings
- Collected works
- Monographs

- NASA reports
- NASA contractor report
- Reports from U.S. and non-U.S. government agencies
- NASA-owned patents and patent applications

The Database contains over 1.6 million records with 50% originating outside the United States. Over 10% of the Database relates to geosciences including earth resources and environmental pollution. The Database is available on-line in the U.S. via Dialog Information Services. The power of on-line access permits fast, precise retrieval from a wide variety of points within a database record.

Roadmaps to the world's literature, such as the Aerospace Database, are of little use, however, if they are not linked to a comprehensive source for the full text of items that they reference. For the Aerospace Database this source is the AIAA Library, the world's largest private collection of aerospace and related material. The Library's holdings include: 1,600 journals, 35,000 books, 800,000 conference and meeting papers, and 850,000 microfiche. These extensive holdings allow the Library to supply full-text copies of items referenced within the Aerospace Database.

Based upon experience within the engineering disciplines, we have identified the following information needs: access, use and an effective interface.

Access to information should not necessarily require those in need to be information retrieval experts. We are seeing an increase in menu-driven systems and CDROM (Compact Disc Read Only Memory) applications that are facilitating the access process.

Our experience at AIAA indicates that, although 50% of the Database's content originates outside the U.S., there is little acceptance of non-U.S. material by U.S. engineers and scientists. While recognizing the problem of dealing with documents in languages other than one's native tongue, non-U.S. material certainly has value. NASA has recognized this value by the establishment of document exchange agreements with foreign organizations that ensure U.S. scientists access to international research. As information providers we need to promote use of this valuable component of our information resources.

The most powerful interaction in information transfer is an effective interface between the individual in need of information and the librarian or information gatekeeper. The combination of these two spheres of professional knowledge linked to an interactive on-line information retrieval system will result in search output of great relevance. By being a part of the process the scientist can suggest additional search terms and directions, while the information professional is ideally equipped to propose resources that cover the subject matter.

New technologies and applications for those technologies with regard to information retrieval systems are constantly being developed. Combine them with the research being conducted in issues of public health and they can lead science to new and important discoveries and solutions to the changing global atmospheric problems facing all of mankind. Our role as information providers is to facilitate this process.

AEROSPACE DATABASE
DIALOG® INFORMATION RETRIEVAL SERVICE

FILE DESCRIPTION

The AEROSPACE DATABASE is the online version of two printed publications: *International Aerospace Abstracts* (IAA) and *Scientific and Technical Aerospace Reports* (STAR). It is a comprehensive engineering and technology information resource providing worldwide bibliographic coverage of published and unpublished scientific and technical literature.

Summary abstracts are provided for over ninety percent of the records from 1972 forward.

SUBJECT COVERAGE

AEROSPACE DATABASE coverage concentrates on all aspects of aerospace research and development, the support of basic and applied research, and technological application to areas such as:

- Chemistry and Chemical Engineering
- Aircraft Design and Instrumentation
- Aerodynamics
- Communications and Navigation
- Space Sciences
- Spacecraft Design and Systems Engineering
- Propellants and Fuels
- Lasers and Masers
- Mechanical Engineering
- Structural Mechanics
- Electronics and Electrical Engineering
- Fluid Mechanics and Heat Transfer

- Quality Assurance and Reliability
- Mathematical and Computer Sciences
- Physics: Solid State, Thermodynamics, Atomic and Molecular, Nuclear and High-Energy, Optics, Acoustics, Plasmas
- Life Sciences
- Geophysics and Earth Resources
- Meteorology, Climatology, and Oceanography
- Environmental Pollution
- Energy Production and Conversion
- Social Sciences

SOURCES

Over 1,600 periodicals from world-wide sources are scanned. Coverage includes journal articles, conferences, books, theses, and unpublished report literature. Approximately fifty percent of the documents originate outside the U.S.

DIALOG FILE DATA

Inclusive Dates: 1962 to the present
Update Frequency: Twice a month (approximately 2,700 records per update)
File Size: Over 1,390,000 records as of October 1985

ORIGIN

The AEROSPACE DATABASE is co-produced by the American Institute of Aeronautics and Astronautics, Technical Information Service (AIAA/TIS) and the National Aeronautics and Space Administration, Scientific and Technical Information Branch (NASA/STIB). Questions concerning file content should be directed to:

Aerospace Database Services   Telephone: 212/582-4901
American Institute of Aeronautics and Astronautics   DIALMAIL: 9176
555 West 57th Street
New York, NY 10019

FILE 108

AEROSPACE DATABASE
DIALOG FILE 108

SAMPLE RECORD

DIALOG Accession Number

```
           1283757    A84-21257
AN=        Modal density of honeycomb plates                                          /TI
CS=        CLARKSON, B. L. (Southampton, University, Southampton, England);
AU=        RANKY, M. F.                                                               SN=
JN=        Journal of Sound and Vibration (ISSN 0022-460X), vol. 91, Nov. 8,
SO=        1983, P. 103-118. Sponsorship: European Space Agency.                      PY=
           Publication Date: Nov. 1983        Refs.
PD=        Contract No.: ESA-4100/79-NL/PP                                            CN=
LA=        Language: English
CO=        Country of Origin: United Kingdom        Country of Publication: United    CP=
           Kingdom
DT=        Document Type: JOURNAL ARTICLE
           Most documents available from AIAA Technical Library
JA=        Journal Announcement: IAA8407
           Honeycomb plates are much stiffer than uniform plates of similar
           mass and consequently their modal density is relatively low. This
           report describes an experimental study of the modal density of uni-
           form honeycomb plates undertaken to verify the theoretical results
           available in the literature. This work was then extended to determine
           the effort of a large cut-out in the center of the honeycomb panel
           typical of a spacecraft platform. It was confirmed that the area of       /AB
           the panel is the controlling parameter. Finally, the effect of addit-
           ional masses, inserts and edge stiffeners was studied. A procedure
           for allowing for these components was developed and confirmed by ex-
           periment. When corrections are made for these effects the theory gives
           a reasonable estimate of the broad band frequency average (500 Hz
           bandwidths are usually required) modal density of typical spacecraft
           components such as platforms and side panels. (Author)
SF=        Source of Abstract/Subtile: AIAA/TIS
           Descriptors: *HONEYCOMB STRUCTURES; *MODAL RESPONSE; *PLATES
           STRUCTURAL MEMBERS); *SANDWICH STRUCTURES; *SPACECRAFT STRUCTURES;         /DE
           HOLES; HONEYCOMB CORES; IMPEDANCE; MANOTS (ESA); MASS DISTRIBUTION;
           PERFORATED PLATES
           Subject Classification: 7539  Structural Mechanics (1975-)                 SC=
```

SEARCH OPTIONS

BASIC INDEX

SUFFIX+	FIELD NAME	INDEXING	SELECT EXAMPLES
/AB	Abstract[1]	Word	S SPACECRAFT(W)PLATFORM/AB
/DE	Descriptor[2]	Word & Phrase	S HONEYCOMB/DE
			S PERFORATED PLATES/DE
/TI	Title[3]	Word	S MODAL(W)DENSITY/TI

+If no suffix is specified all Basic Index fields are searched. [3]For most 1962 and 1963 records, titles are included in
[1]Present in over 90% of records from 1972 forward. in the Source (SO=) field. Supplementary title
[2]Also /DF. information may also be included in the Title field.

ADDITIONAL INDEXES

PREFIX	FIELD NAME	INDEXING	SELECT EXAMPLES
AN=	Accession Number	Phrase	S AN=A84-21257
AU=	Author	Phrase	S AU=CLARKSON, B. L.
AV=	Availability	Word	S AV=AIAA
BN=	International Standard Book Number (ISBN)	Phrase	S BN=0-309-03380-2
CN=	Contract Number	Word & Phrase	S CN=ESA
			S CN=ESA-4100/79-NL/PP
CO=	Country of Origin	Phrase	S CO=UNITED KINGDOM
CP=	Country of Publication	Phrase	S CP=UNITED KINGDOM
CS=	Corporate Source	Word	S CS=(SOUTHAMPTON(W)UNIV?)
CS=	Corporate Source Code	Phrase	S CS=U8734375
DT=	Document Type	Phrase	S DT=JOURNAL ARTICLE
JA=	Journal Announcement	Phrase	S JA=IAA8407
			S JA=IAA
JN=	Journal Name	Phrase	S JN=JOURNAL OF SOUND 'AND' VIBRATION
LA=	Language	Word	S LA=FRENCH
NT=	Notes	Word	S NT=SPONSORED
PD=	Publication Date	Phrase	S PD=8311
PI=	Patent Information	Phrase	S PI=US-PATENT-APPL-SN-243683
PU=	Publisher Data	Word	S PU=(BROOKS(W)AFB)
PY=	Publication Year	Phrase	S PY=1983
RN=	Report Number	Word & Phrase	S RN=NASA
			S RN=NASA-SP-459
SC=	COSATI Codes[4]	Phrase	S SC=6E
SC=	COSATI Code Text[4]	Word & Phrase	S SC=(CLINICAL(W)MEDICINE)
			S SC=CLINICAL MEDICINE?
SC=	Subject Category Code	Phrase	S SC=7539
SC=	Subject Category Text	Word & Phrase	S SC=MECHANICS
			S SC=STRUCTURAL MECHANICS?
SF=	Source of Abstract/Subfile	Phrase	S SF=DISSERT. ABSTR.
SL=	Summary Language	Word	S SL=ENGLISH
SN=	International Standard Serial Number (ISSN)	Phrase	S SN=0022-460X
SO=	Source	Word	S SO=(EUROPEAN(W)SPACE)
SP=	Sponsoring Organization Code	Phrase	S SP=T4425772
UD=	Update	Phrase	S UD=9999

[4]Present for about half of the STAR records.

LIMITING

Sets may be limited by Basic Index suffixes, i.e.,
/AB, /DE, /DF, and /TI, as well as by the features
listed below:

SUFFIX	FIELD NAME	EXAMPLES
None	Publication Year	S S3/1984
/ENG	English Language	S S1/ENG
/NONENG	Non-English Language	S S8/NONENG

SORTING

SORTABLE FIELDS	EXAMPLES
Online (SORT) and offline (PRINT). AN, AU, CS, JN, PD, PY, SN, TI.	SORT 3/ALL/PD,D PRINT 8/5/ALL/PD

FORMAT OPTIONS

NUMBER	RECORD CONTENT	NUMBER	RECORD CONTENT
Format 1	DIALOG Accession Number	Format 5	Full Record[1]
Format 2	Full Record except Abstract	Format 6	Title and DIALOG Accession Number
Format 3	Bibliographic Citation	Format 7	Bibliographic Citation and Abstract[1]
Format 4	Full Record[1] with Tagged Fields	Format 8	Title and Indexing

DIRECT RECORD ACCESS

FIELD NAME	EXAMPLES		
DIALOG Accession Number	TYPE 1344408/5	DISPLAY 1331202/4	PRINT 1329895/5

108-2 (Revised October 1985)

References

Dialog Information Services
3460 Hillview Avenue
Palo Alto, CA 94304
1-800-3-DIALOG

Directory of Directories
Gale Research Company
Book Tower
Detroit, MI 48226

Directory of Online Databases
Cuadra/Elsevier
655 Avenue of the Americas
New York, NY 10010
212/989-5800

Encyclopedia of Associations
Gale Research Inc.
Book Tower
Detroit, MI 48226
313/961-2242

Going Online
An Introduction to the World of Online Information
Learned Information, Inc.
143 Old Marlton Pike
Medford, NJ 08055
609/654-6266

NASA Scientific and Technical Publications
NASA
Scientific and Technical Information Division
Washington, DC 20546

Additional Sources

Consultants and Consulting Organizations Directory
Editorial Services Limited
P.O. Box 6789
Silver Springs, MD 20906
301/871-5280

Datapro Directory of On-Line Service
1805 Underwood Blvd.
Delran, NJ 08075
609/764-0100

Directory of Fee Based Information Services
Burwell Enterprises
5106 FM 1960 West, Suite 359
Houston, TX 77069
713/537-9051

Directory for Special Libraries and Information Centers
Gale Research Inc.
Book Tower
Detroit, MI 48226
313/961-2242

PANEL DISCUSSION: INFORMATION SOURCES AND NEEDS

Moderator:

Marta Dosa, Professor, School of Information Studies, Syracuse
University

Panelists:

Gerald S. Barton, Director's Staff, National Oceanographic Data Center,
NOAA
E. Joseph Bangiolo, Chief, Evaluation Sector, Information Technology
Branch, National Institute of Allergy and Infectious Diseases
Geoff Worton, Director, Professional Services, American Institute
of Aeronautics and Astronautics, Technical Information Service
Donna Orti, Program Manager, Health Education Program, Agency
for Toxic Substances and Disease Registry

DR. DOSA: As usual with interdisciplinary groups, I immediately feel a great rapport with you and I look forward to a very informal and interesting session. I was especially looking forward to this session because, in my experience, the more interdisciplinary a group is, the more interaction we have, a feeling of a common bond. Why would we be here if not for a common goal? I feel that we do have a very special bond here, not the gloom, not the doom, not the fear of the consequences of global climate change, but a very constructive, very positive feeling that, I believe, we all share. It is a sense that decision making and the solving of problems can be improved and that information, data and knowledge are the keys.

I see this session with the theme of "information sources and needs" as having several distinct characteristics. It's very international, not only because of the well-known attribute of "global" climate change, but also because of the specific meaning of the consequences of climate change, not only for the high technology countries, but also for developing countries where I have been working. This is the mission to which colleagues in the third world respond, the theme around which we can really build cooperation. I also feel that this theme demonstrates the need, and our awareness of the need, to put research into action.

And I feel a very high level of public participation, knowing that people may think of this topic as distant and so long-range that the concerns are not of immediate significance. But my experience is that, if they have the slightest encouragement for action, the motivation is there.

I would like to set a context, a framework, for our session with the words of the late scientist Jacob Bronowski. These words have been with me for many years. He wrote that there is no way to exchange information without exercising an act of judgment. Information and data have to be credible, have to have integrity, have to have validity. They must come from a source that is respected and has a very high level of nonformal authority and scientific validity. I feel, in accord with Bronowski's words, that we all exercise and act on judgment whenever we offer and use information and data or informally exchange them.

MR. BARTON: The interagency working group is very concerned about the cost issue, and the approach being taken is to try to implement policy so that researchers can have access to data in anybody's agency. For the Earth Observing System (EOS) data management and the EOS effort, that $150 million or $300 million, or whatever the figure is, is just astronomical. Data management is the key to that program from the start and about 50 percent of the budget goes into data management. That includes getting data, making it available to any researcher in NASA and maybe in the global change community, and then making sure that it is documented and transferred to permanent archives after a period of time.

A major portion, or at least compared to what used to be, is going for data management, including budgeted money to offset the technology changes through time. It is budgeted so that every five years, if need be, new equipment can be purchased to keep up with the technology and we are not caught in a bind of changing from Compact Disc Read Only Memory (CDROM) as we're using now to some new technology.

MR. BANGIOLO: I just want to say, pay as you go may make sense in terms of some objectives, but I'll be darned if I can figure out what they are. In terms of the long-term objectives, it doesn't make any sense at all.

MR. WORTON: I share the same concerns about control vocabularies across databases and duplication that exists between databases. Dialog introduced a single Read Only Memory (ROM) search which allows you to address a variety of different databases with the same search strategy, which sounds like a wonderful idea, except it tends to highlight the level of duplication that exists between databases in a particular discipline. It's not much fun getting the same record three times and paying for it three times. It also highlights the inconsistencies in terms of application of controlled vocabulary and terms. So that, while the terms that you've used or the strategies that you've used might work exceptionally well in one database, it might not apply as well to others.

MR. BARTON: Could you provide a bit more information on the status of the NASA Scientific and Technical Information (STI) system, how people access it and what the future of it is? The NASA RECON System [a bibliographic database maintained by NASA] is, if not the first, one of the first on-line information retrieval systems. It was developed under a contract by Lockheed. Dialog Information Services, which used to be a Lockheed subsidiary, grew out of the good work that Lockheed did at NASA RECON. It has a variety of databases, some of which are classified. Access to it is restricted, I think, to government contractors, to NASA Centers and NASA contractors and, within that access, there are different degrees of classification. If you have access to our database on RECON, there are some other databases that you can't get.

I know that one of the things that they're addressing at the NASA STI Division is what they can do about providing access to some of the databases that are developed at various NASA centers around the country. The NASA RECON System is located in Baltimore-Washington International Airport in Maryland. But there will be databases at Langley and NASA Goddard and Ames that won't be on that system, but could be of value to the people that have access to that system. That's a problem for them to address.

The RECON System, as I understand it, is a directory to NASA literature and gray literature at NASA. I don't know if this morning anybody mentioned the gray literature that exists at the National Technical Information Service (NTIS). NTIS has databases that are like the NEDRAS database on DIALOG so that you can access these. There are bibliographic references to government gray literature, a wealthy source of information that you can't find any other way unless you know a particular person in an agency that generated this kind of a report. Those reports do go to NTIS, they do get catalogued, and you can search them and obtain them through NTIS as a buying service.

DR. DOSA: Do you mean that desktop published material would also be included?

MR. BARTON: Yes, it could be, if it's submitted.

ROBERT SHOPE (Yale University School of Medicine): I want to direct some comments to Ms. Orti. I thought her presentation was on the mark. You've highlighted some of the things that need to be done and some of the problems. There are a couple of problems that we deal with: one is getting the information to the physicians and the other is getting information from the physician back to the patient. And, Ms. Orti, you said that there were no courses in environmental health in the medical schools—that's almost true. We do have courses at the Yale School of Medicine. They're elective and the problem isn't with the courses; the problem is getting the students to elect them. If you try to get something in the medical school curriculum which is required for the students, you're competing with all sorts of other disciplines and public health doesn't compete too well. Most medical schools are not oriented to public health; they're oriented to treatment of patients and that's a problem I don't know how to get around. After medical school, there are some possible ways of getting information. There is a TV channel on cable TV which is aimed at physicians. I have several colleagues who listen to it and watch it. That's one possible way.

To get the information from the physician to the patient is even more difficult because contact with the patient is very short. I don't have figures, but I know from personal experience that most of the time spent involves the patient telling the physician what his or her problems are, not the other way around. Information transfer from the physician to the patient is, at best, difficult. Again, one could use the media if you could get physicians to appear on talk shows either on television or radio. I hope you keep trying and I wish you success.

MS. ORTI: I appreciate your comments. We have another effort called Educating Physicians in Occupational Environmental Health. We're working with Duke University to introduce curriculum to the medical schools and they have a network of five medical schools which will eventually teach the ten southern states. We are looking at getting environmental issues into the medical school, but it's a rocky road because, frankly, medical students and physicians have many other concerns and environmental concerns generally are not on the top of the list. It's not just in medical schools, it's in elementary and high schools throughout the country. Environmental concerns are simply not addressed in science classes and it's an issue that probably needs to be developed by some groups. How can we bring environmental issues to the attention of the public and make those programs informative and dependable, giving good, reliable information, instead of scaring people or giving them information which makes them skeptical about the whole issue?

MICHAEL P. FARRELL (Oak Ridge National Laboratory): It's actually worse than you said it is. We just reviewed the high school textbooks on climate change and the physics of climatology. There's a textbook that has nine Nobel Prize winners on the cover and the text was incorrect in terms of physics. One of the Nobel prize winners whom I know said he'd never seen or read the book and he doesn't know how his name got on the cover.

DR. DOSA: Misinformation is terrible when it's political but, when it's scientific, it's just as dangerous or more so.

JACQUELINE TROLLEY (Institute for Scientific Information): I had intended to come down here this morning just as a market and research trip, but I'd like to share with you some information on the product that we're developing that might be of help to a lot of you. Early in 1990, we will be coming out with a new diskette product that has no print counterpart. It's going to be called Focus on Global Change and it will be published on a bi-monthly basis. It will be international in scope and we're running it against 8,000 of our journals, our entire database. It's something quite new for us. Most of our products are either geared toward one particular discipline, for instance the Science Citation Index or maybe Current Contents, where we have several group disciplines. This new product will cover our social sciences, arts and humanities, and science journals. We have also had to add, for this particular product, a number of new journals that we have not covered in the past. Typically the Institute for Scientific Information (ISI) does not cover a journal until it has a publishing history. We're making an exception for this product. We have brought in some of the new things, such as newsletters that are coming out in this particular area. We're pretty excited about it. We are very sure that it will be easy to use. Someone has talked this morning about the idiot-proof approach to on-line services and, in the absence of that, with the development of Current Contents we were able to come pretty close to an idiot-proof product. It's menu driven. This product will also be menu driven. It will be on the article level, as opposed to our Current Contents products which are on the journal level. We will have, in this edition, the bibliographic information only.

STATEMENT FROM THE FLOOR: What does that mean? I don't understand your terminology. What do you mean by "on the article level"?

MS. TROLLEY: With Current Contents, which has been around for 30 or 35 years, we take the contents pages of the journals. This time we are not taking an entire journal issue and listing out the contents pages. We're going right into the articles. We're running essentially a gigantic profile against our 8,000 journals and the database and working it that way. We're just taking core articles out, instead of trying to find journals that are specifically related to this particular subject area.

FREDERICK W. STOSS (Center for Environmental Information): Many of the current database producers are missing a percentage of the fugitive literature. That percentage may vary from topic to topic, but the work that we've done on acid rain for six years has indicated that some of the mainstay scientific, policy and social science databases are missing — perhaps 20 percent of the information in some broad categories. State agencies are doing a considerable amount of work in acid rain research. Nonprofit research centers and industries are producing technical reports some of which, when you go into Science Citation Index, turn up very strong citation patterns. They may have been setting some precedent-oriented research. Two exceptions which immediately come to mind on DIALOG are the Electric Power Database produced by the Electric Power Research Institute and Little On-Line which is the report literature from Arthur D. Little. They've circumvented the problem of fugitive literature by developing their own database and making it available to those who use that particular service. One of the things that's been intriguing from a scientific information perspective has been dealing with the traditional way that scientific information is generated. In any textbook on sci-tech information, you run into a little diagram where a scientist gets an idea. Then you go through a record or a transition process of how information is produced as you go from that idea to a laboratory notebook, to a conference proceedings or preliminary publication to a full-fledged peer reviewed publication or technical report. Then it goes to secondary sources, such as textbooks, and the cycle becomes complete when another scientist reads about it and starts another paper based on the initial research. This cycle shows how information begets more information.

We're seeing a trend in the information arena where trying to find a faster way to disseminate information necessitates the institutional and therefore the financial support to continue research in those areas. Take a look at the research fronts on acid rain or global climate change in ISI's products. You'll soon see that there's even an evolution that can be tracked there and perhaps we're even trying to circumvent the peer review process. Those of you who have submitted papers or serve as peer reviewers for journals know that it can sometimes be a long and tedious process, and the information is needed yesterday. We're seeing a rush of desktop publishing and the creation of newsletters. We're seeing, almost daily, new global climate change publications being produced which are disseminating information faster than we've ever seen before. As more organizations pick up on the desktop publishing technologies and more people start to dump things into CDROM products, and as databases become more accessible to a much wider community, keeping track of that information is going to become a much more difficult challenge.

We're seeing a diffusion of the vehicles. In the past they have been very tightly controlled, and that's why we see a lack of index control on databases. Sid Siegel, in 1978 or so, was developing the Chemical Substance Information Network (CSIN) program, the only practical application of using a single query to a database which would translate into about 14 or 15 other indexing languages and dump out one search that didn't have any duplicates. I think it was the GENTOX (GENetic TOXicology) program. Whatever happened to some of the technologies that went into developing CSIN which linked 20-odd different components, such as bibliographic databases and a lot of this fugitive literature? We are starting to outrace our ability to control some of these technologies. And we're going in the opposite direction from what we are striving for; we're diffusing information faster than we can control it. Does anybody want to tackle any of these things? I don't know what this is doing other than creating frustration for those of us who are trying to locate and centralize information.

MR. BARTON: I could look at what could be considered fugitive data sets. The major databases and major directories that I've talked about — NASA, NOAA and the USGS — would contain mainly internal data sets, data that would be useful for NASA people that are generated by NASA or by NASA contractors and data sets that we have in NOAA generated by NOAA principal investigators that are archived in the data center or perhaps developed by contractors. What we did in the NEDRAS was to not limit data descriptions to specifically NOAA data sets. We include everything. We have data sets from very obscure locations like the Calvert Cliffs power plant down on the Chesapeake Bay. These people have been recording data, per EPA and DOE requirements, for years and years. There's meteorological, weather and water quality data and some biological kinds of data, long-time series data with a lot of information on the Chesapeake Bay. Many of those kinds of data sets are documented in NEDRAS. What particularly struck me was this older part that I described before, the Index. There were data sets described that somebody at a SUNY University had. There were about five or six elephant studies that were done in Africa, I believe dating back to the mid-60s.

DR. DOSA: There's one at Cornell right now.

MR. BARTON: There are a variety of things like that, that nobody else is worrying about. They're very difficult to locate. I don't know what's going to happen with NEDRAS in the future, but these fugitive kinds of data sets represent the true value of it. But the problem exists in literature also.

DR. DOSA: Let me share with you my own feeling that almost all topics have at least been touched upon, if not covered. What I didn't hear mentioned, and would like to leave you with, is a hint about the legal aspects of our information transfer. As we are talking in international terms, I remembered the recent developments with the International Telecommunications Union and the World Telephone and Telegraph Conference where the voting power of the developing countries have come across as a loud and clear message. As we're becoming more and more global, as our private enterprises globalize, and as management schools teach more and more international aspects, our need for understanding the legal implications of our reaching out increases. Two scenarios will help me bring this to a close. One scenario is from the point of view of the North and of the information-affluent part of the world, for example Japan, closing the fifth generation project in 1992 and already beginning the sixth generation project on the biological computer, Europe with the unified market in 1992, and the United States. This three-way configuration is going to be joined by the third world information needs and information power in many senses.

But this information power on the part of the developing countries is still rooted in a cultural scenario. A few years ago I heard the statement, "In Africa, when an old man dies, a library burns down." The illiterate societies are still realities. Information and data are cultural commodities and require culture-based processes. We move at two levels, information handlers and information users. One level is the legal, policy and technological level. The other level is cultural, professional and person-to-person cooperation. Professional reaching out is a tremendous influence against political and legal problems. So I would like to thank you all for giving me a sense of this unity this morning. Thank you.

INFORMATION TRENDS AND RESOURCES

Frederick W. Stoss

Air Resources Information Clearinghouse (ARIC)
Center for Environmental Information, Inc.
99 Court Street
Rochester, New York 14604

Introduction

Bibliometric analysis has been used to track the growth of information on the topic of acidic deposition (Stoss, 1988; Lancaster and Lee, 1985). Such tracking can be useful in the identification of topics with scientific or policy significance. For environmental topics that are driven by strong policy developments (e.g. acid rain, global warming, hazardous waste disposal, ozone depletion), this tracking procedure can be useful for staying abreast of rapidly changing events and developments.

Many of the scientific and general information indexes and abstracts are available in electronic format and can provide a rapid and inexpensive means to analyze the generation of information on broad or specific topics. Examination of the publishing history on a specific topic can reveal the emergence of that topic as a special area of concern.

Methods

Bibliographic databases covering general science and technology (SCISEARCH), energy and environmental science technology (DOE ENERGY), general information (MAGAZINE INDEX and AP NEWS), and the medical or health care literatures (EMBASE, MEDLINE)[1] were used to search for information related to the topic of climate change. A general description of these databases is provided as an appendix to this paper.

---

1 All bibliographic databases were searched on DIALOG, Dialog Information
  Services, Inc., a Knight-Ridder company.

Published 1990 by Elsevier Science Publishing Co., Inc.
Global Atmospheric Change and Public Health
James C. White, Editor

The topic of climate change[2] is one that encompasses the two major concepts of "global warming" and "ozone depletion." The concept of global warming resulting from the accumulation of carbon dioxide and other atmospheric gases and the entrapment of solar radiation in the lower atmosphere has been known since the mid-1800s (Arrhenius, 1896). Ozone depletion, on the other hand, is a concept of more recent origin (Molina and Rowland, 1974; Rowland and Molina, 1975) and received considerable attention in 1985 when United Kingdom scientists reported a dramatic seasonal thinning of the ozone layer over the Antarctic (Farman, Gardiner and Shanklin, 1985).

These concepts and key words were examined in the selected databases through Dialog's DIALINDEX file, where only the index terms, key words, title words, and other searchable fields are directly accessed. The output from a DIALINDEX search is a simple list of the frequency of the terms that can be retrieved from the databases selected. Boolean and phrase searching is possible in DIALINDEX, which allows for a more refined manipulation of the bibliographic data (e.g. analyzing the output by year of pub-lication).

Results and Discussion

The growth of the topic of climate change in the selected databases is plotted from 1973 to 1989 and is shown in Figure 1. It should be noted that the number of citations for the year 1989 are lower than the actual number published, because of the lag-time in having the bibliographic information compiled for the on-line databases. A decrease, therefore, in the number of citations retrieved in 1989 cannot be construed as being a reduction in the production of information for that year.

A more refined analysis of this information growth was performed for the scientific literature (SCISEARCH) and energy literature (DOE ENERGY). Figures 2a and 2b show the topic break-out of global warming and ozone depletion topics in these two databases. The more rapid emergence of climate change in the scientific literature (SCISEARCH) is shown to occur between the years of 1983 and 1984. The analysis of bibliographic information from the DOE ENERGY database, on the other hand, indi-cates that there has been a steady increase in the growth of literature on global warming for more than one decade, with the issue of ozone depletion showing dramatic increases in recent years. The differences also reflect the different coverage of the literature by the two database producers and demonstrates the need for conducting multi-database searching for providing comprehensive coverage of a topic.

---

2   Key words/concepts: "Global Warming:" climate change, greenhouse effect/gases, global warming, ocean warming, sea level rise, warming climate; "Ozone Depletion:" Antarctic/Antarctica, chlorofluorocarbon/CFCs — emissions/ monitoring/substitutes, ozone, ozone depletion, ozone hole, ozone layer, stratospheric ozone.

The emergence of this same topic as an environmental topic occurs much later (1985-1988) in the more general magazine and newspaper literatures. The development of global warming as a "hot" environmental topic and the dramatic emergence of ozone depletion as an issue-oriented topic is clearly demonstrated in their publication record in MAGAZINE INDEX (Figure 2c). The fact that the past several years have been highlighted by dramatic increases in regional warming and drought has enhanced this effect in the general print media.

If we examine the growth of information on this same topic in several medical databases,[3] we can attempt to determine the extent to which the medical professions are examining or discussing specific health effects (direct or indirect) in response to changing climates. The results of searching the medical literature from 1980 to 1989 are presented in Figure 3.

Citations retrieved from the HEALTH PERIODICALS DATABASE reflect a general introduction of the topic similar to that displayed in MAGAZINE INDEX. Of the 175 citations retrieved from the HEALTH PERIODICALS DATABASE, a total of 15 were categorized as being relevant to the direct or indirect medical or health aspects of the topic (the major areas of research being the impact of increased ultraviolet radiation to the formation of skin cancers, and the nutrition or infectious disease ramifications of global warming; other "medical" articles tend to be more speculative in their description of potential impacts to human health). All other citations were of a very general nature, and serve primarily as introductions of the topics of climate change to the readership.

The information base related to climate change in the medical literature does not reflect a large scientific or technical output of a research effort directed specifically to the impacts of climate change to human health. EMBASE has, by far, the largest retrieval rate of citations on this topic. From 1974 to November of 1989 a total of 459 citations on the general topic of "climate change" are found; 258 of the citations were retrieved between 1982 and 1989. Of these articles only 43 or so are describing specific medical conditions that result from some aspect of climate change. While the nature of the topic is not new, its emergence as a critical topic of concern is only beginning to make its way into the medical literature.

A bibliography of selected titles retrieved from the three medical databases was compiled and is appended to this paper as a representation of the medical literature related to the topic of climate change.

---

3   HEALTH PERIODICALS DATABASE, MEDLINE and EMBASE (all
     available through Dialog Information Services).

Summary and Conclusions

The concepts of climate change were tracked in the general, scientific and medical literatures over extended periods of time. The results show an increase in the yearly production of information related to the topics of "global warming" and "ozone depletion." While these issues emerged as prominent issues in the scientific literature in the early-1980s, it was not until 1987-1988 that the same topics were as dramatically drawn to the attention of the general lay readership or to the medical and health care communities.

The topic of "climate change" has been recently introduced to the medical and health care communities in general terms. As public awareness and policy options related to climate change increase, so should the institutional support for research in these areas increase. As more effort is placed into research on the health effects related to climate change, we would expect to see the medical literature on this topic increase in numbers.

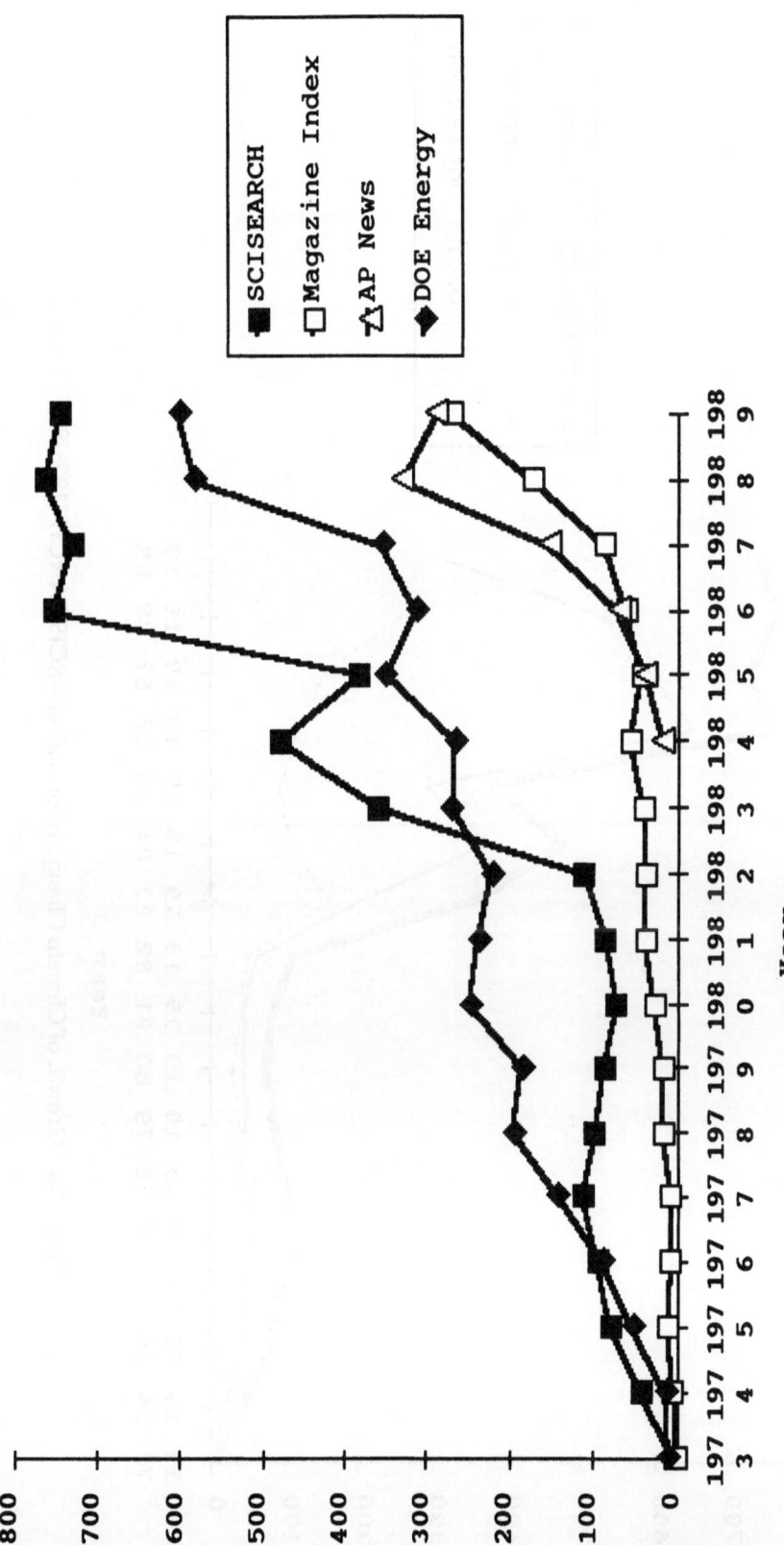

Figure 1. Growth of Climate Change Information in Two Popular Literature
and Two Scientific Databases (1973-89)

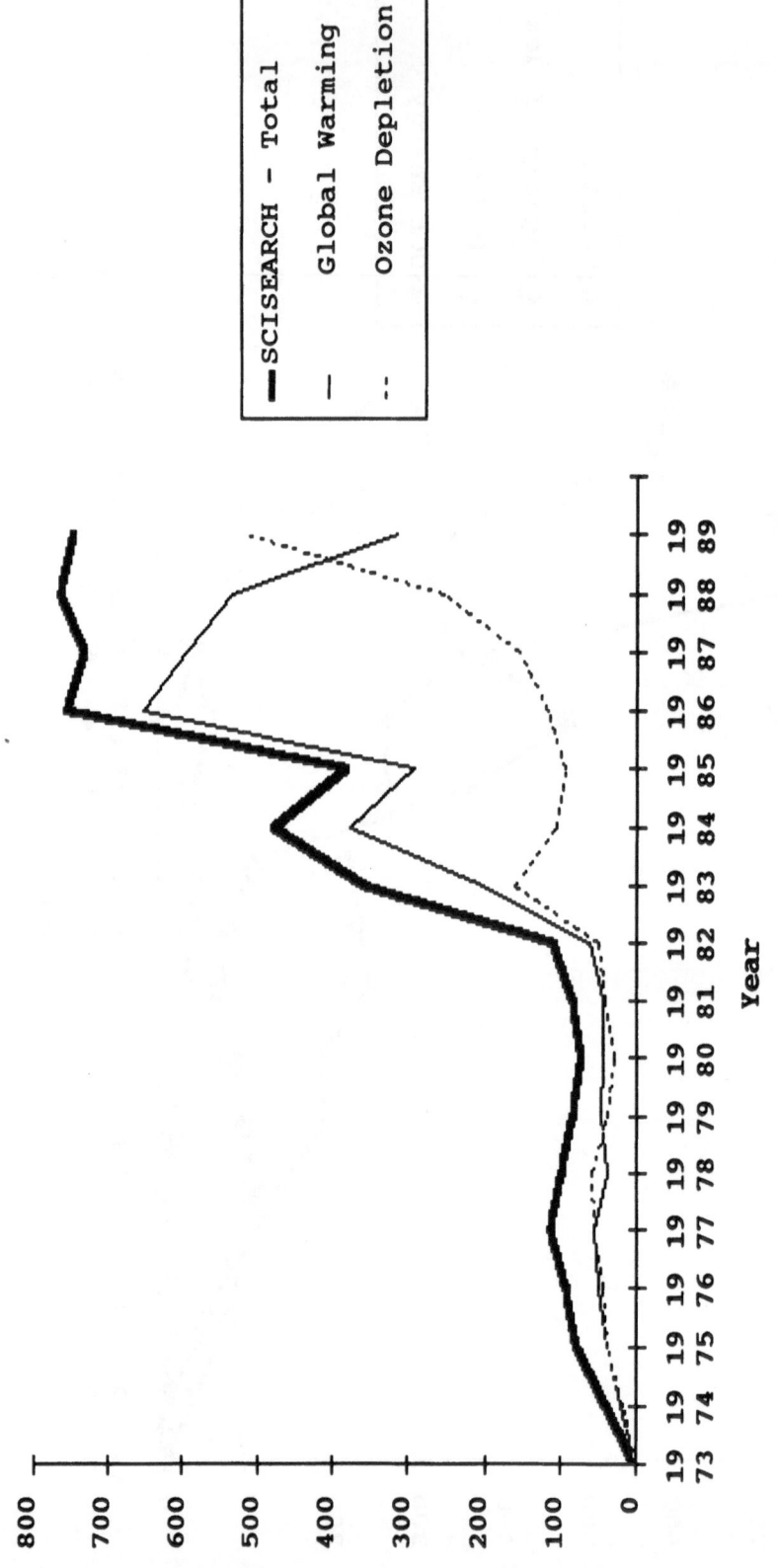

Figure 2a. Growth of Climate Change Information—SCISEARCH (1973-89)

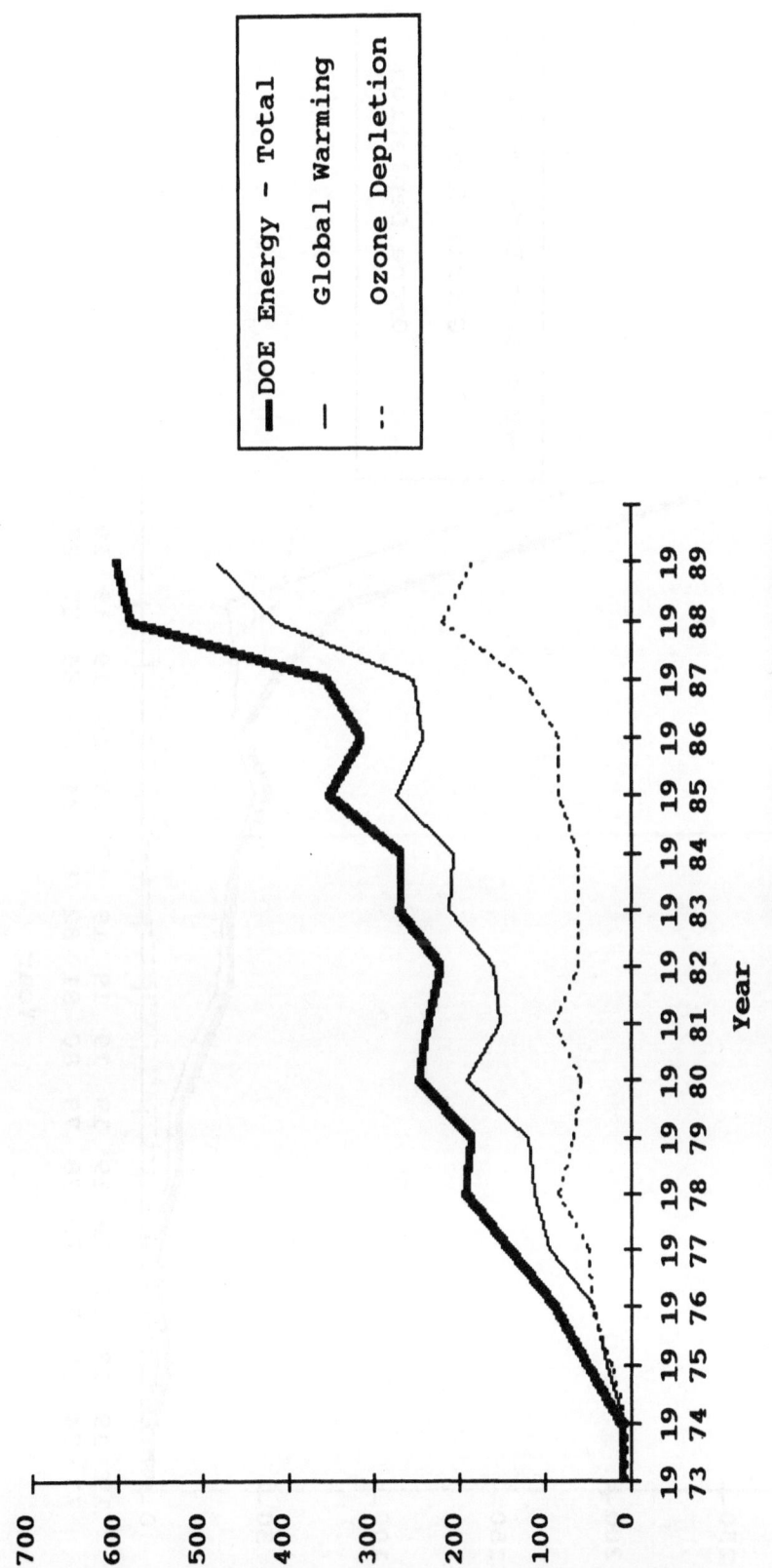

Figure 2b. Growth of Climate Change Information — DOE ENERGY (1973-89)

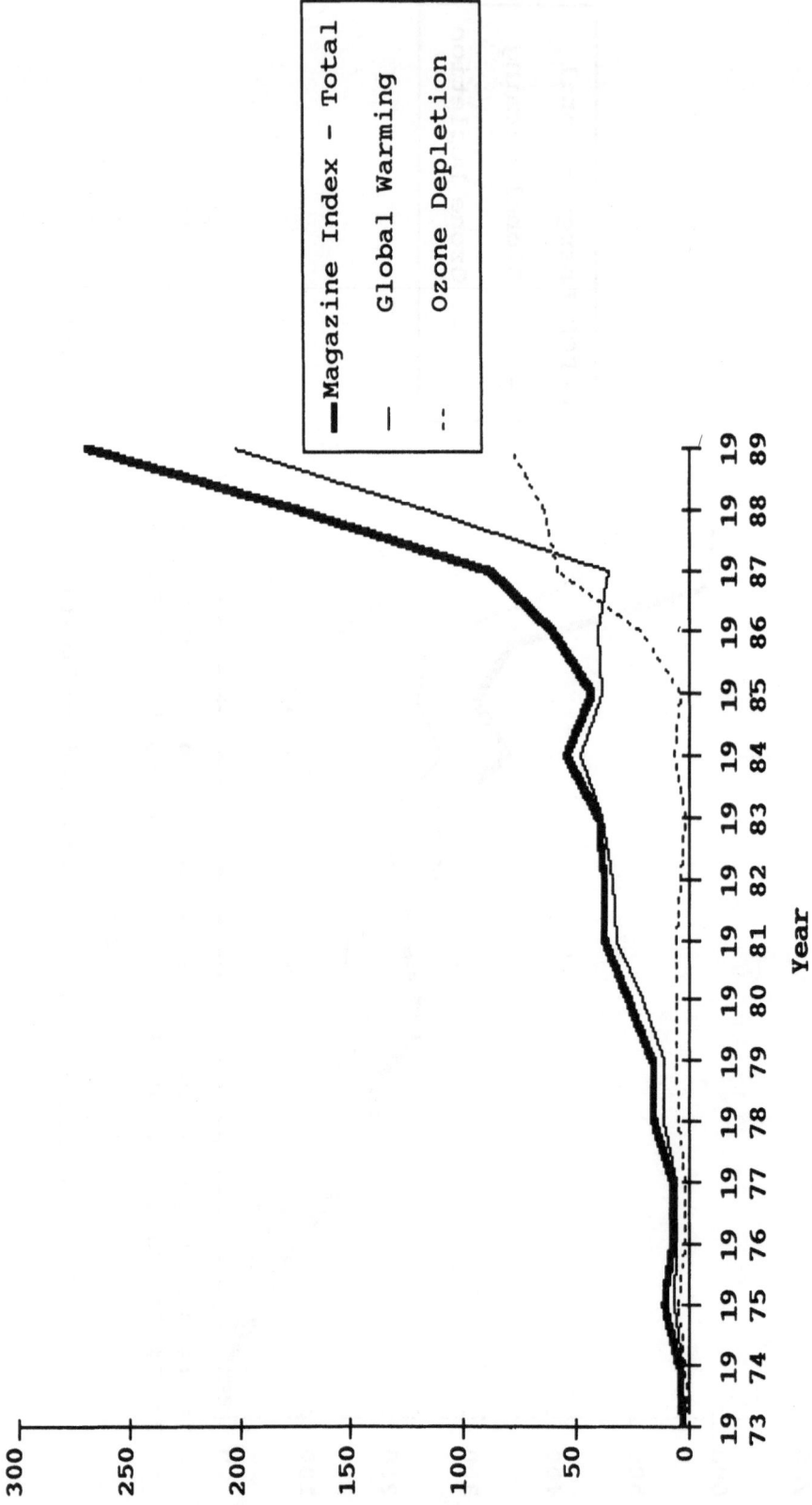

Figure 2c. Growth of Climate Change Information — MAGAZINE INDEX (1973-89)

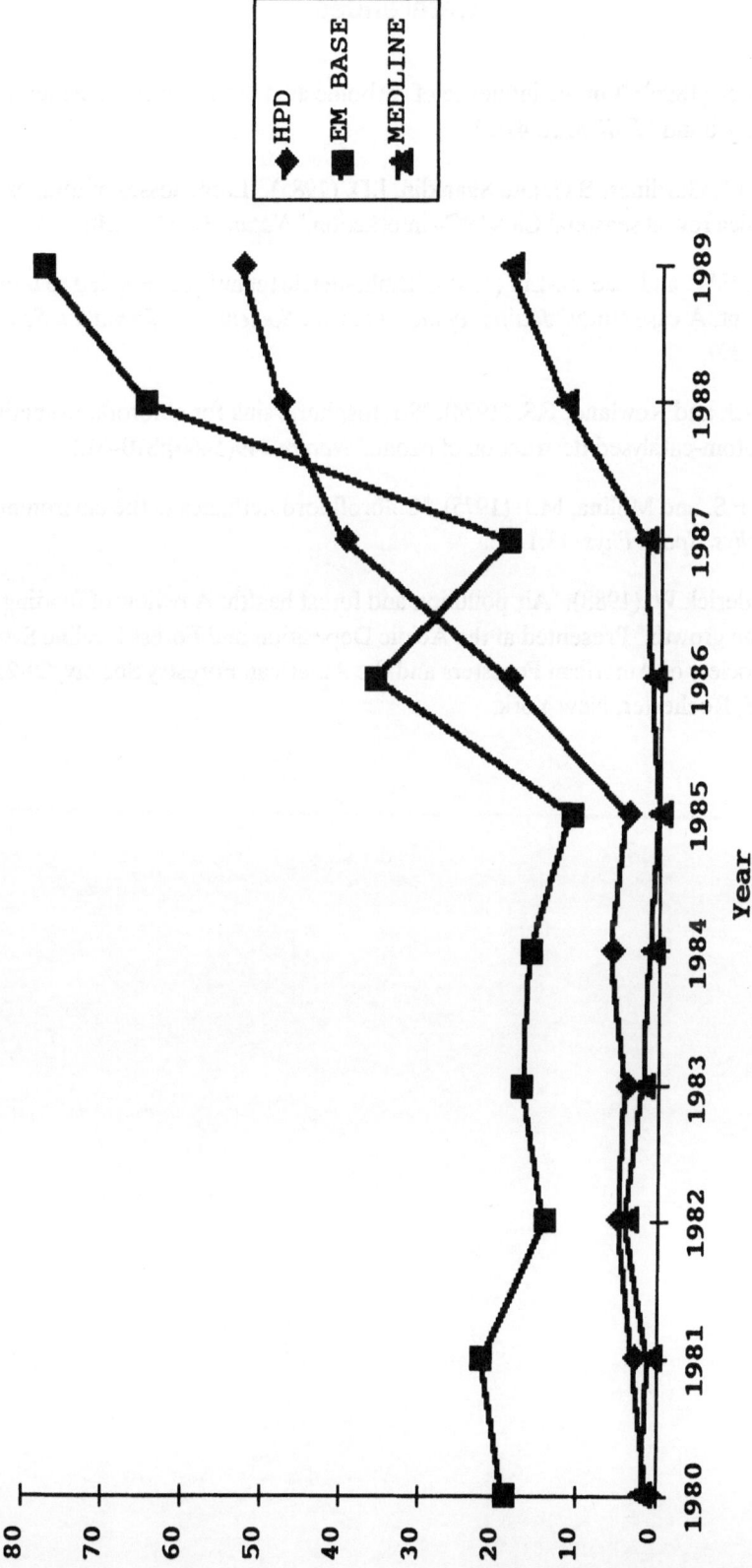

Figure 3. Growth of Climate Change Information in Medical Databases (1980-89)

References

Arrhenius, S. (1896). "On the influence of carbonic acid in the air upon the temperature of the ground." *Phil. Mag.* **41**:237.

Farman, J.C., Gardiner, B.G. and Shanklin, J.D. (1985). "Large losses of total ozone in Antarctica reveal seasonal ClO$_x$/NO$_x$ interaction." *Nature* **315**:207–210.

Lancaster, F.W. and Lee, Ja-Lih (1985). "Bibliometric techniques applied to issues management: A case study." *Journal of the American Society for Information Science* **36**(6):389–397.

Molina, M.J. and Rowland, F.S. (1974). "Stratospheric sink for chlorofluoromethanes: Chlorine atom-catalysed destruction of ozone." *Nature* **249**(5460):810–812.

Rowland, F.S. and Molina, M.J. (1975). "Chlorofluoromethanes in the environment." *Rev. Geophys. Space Phys.* **13**:1–35.

Stoss, Frederick W. (1988). "Air pollution and forest health: A review of funding and information growth." Presented at the Acidic Deposition and Forest Decline Symposium, Society of American Foresters and the American Forestry Society, 20-21 October 1988, Rochester, New York.

Appendix 1.   Databases Selected for Study of the Growth of Information on Climate Change

AP NEWS

AP (Associated Press) NEWS provides the full text of its coverage of the national, international, and business news, as well as sports and financial information. AP NEWS is available 24 hours after the data are transmitted on its high-speed DataStream service. All words within the text are searchable, thus providing access to any subject that has been in the news. The Associated Press is the largest supplier of general-interest news to the media world-wide, serving more than 15,000 newspaper and broadcast outlets in 115 countries around the world. AP NEWS is compiled by more than 1,100 journalists in 141 United States news bureaus and 83 overseas news bureaus. In addition, as a U.S. news cooperative, the AP has access to the news compiled by approximately 1,500 newspaper members and 6,000 radio-television members in the United States.

DOE ENERGY (U.S. Department of Energy)

DOE ENERGY is a multidisciplinary file containing world-wide references to basic and applied scientific and technical research literature. The information is collected for use by government managers, researchers at the national laboratories, and other research efforts sponsored by the Department of Energy and for transfer of the results of this research to the public. Abstracts are included for records from 1976 to the present. The database corresponds in part to Energy Research Abstracts and INIS Atomindex, as well as to other publications.

The information is provided by the U.S. Department of Energy, its contractors, other government agencies, professional societies, and through the International Energy Agency's multilateral information program (Energy Technology Data Exchange) and the International Atomic Energy Agency's International Nuclear Information System (INIS). The file includes references to journal literature, conferences, patents, books, monographs, theses, and engineering and software materials. Approximately 50% of these references are from non-U.S. sources.

EMBASE (Elsevier Science Publishers)

The Excerpta Medica database, EMBASE, is one of the leading sources for searching the biomedical literature. It consists of abstracts and citations of articles from over 4,000 biomedical journals published throughout the world. It covers the entire field of human medicine and related disciplines. The on-line file corresponds to the 43 specialty abstract journals and 2 literature indexes which make up the printed Excerpta Medica, plus an additional 100,000 records annually that do not appear in printed journals. EMBASE provides access to periodical articles from more than 3,500 primary journals from over 100 countries. An additional 1,000 journals are selectively screened for relevant articles.

HEALTH PERIODICALS DATABASE (Information Access Company)

HEALTH PERIODICALS DATABASE provides indexing and full text of journals covering a broad range of health subjects and issues. Subjects covered range from prenatal care to dieting, drug abuse, AIDS, biotechnology, cardiovascular disease, environment, public health, safety, paramedical professions, sports medicine, substance abuse, toxicology, and much more. Articles are collected from core health, fitness and nutrition publications. The database provides a valuable resource for corporate, medical and legal librarians, human resource professionals, and product analysts.

Source publications include more than 240 journals and magazines covering health, medicine, fitness and nutrition, plus publications elsewhere indexed by Information Access Company. Examples of the core publications include: *AIDS Alert, FDA Consumer, Drug Abuse Report, Harvard Medical School Health Letter, Medical SelfCare*, and *Tufts University Diet and Nutrition Letter*. Technical medical journals covered in the file include: *American Heart Journal, JAMA* (The Journal of the American Medical Association), *Lancet*, and the *New England Journal of Medicine*. General interest publications reviewed for health-related articles include: *Changing Times, Health and Society, Newsweek, Science News*, and *Sports Illustrated*.

MAGAZINE INDEX (Information Access Company)

MAGAZINE INDEX is the first on-line database to offer truly broad coverage of general interest magazines. This database was created especially for general reference, to handle a constant flow of diverse requests for information from the scholarly to the lighthearted. MAGAZINE INDEX covers more than 435 popular magazines and provides extensive coverage of current affairs, the performing arts, business, sports, recreation and travel, consumer product evaluations, science and technology, leisure-time activities, and other areas. In addition to general reference, MAGAZINE INDEX will serve business and government libraries with information not available on any other on-line database. Users in the fields of market research, public relations, journalism, food and nutrition, and the social sciences will find MAGAZINE INDEX to be a significant resource. In addition to its extensive indexing, MAGAZINE INDEX contains the full text of records from more than 100 magazines from 1983 to the present.

MAGAZINE INDEX provides coverage of more than 435 of the most popular magazines in North America and includes coverage of all magazines indexed by the Reader's Guide to Periodical Literature. The full text of records from more than 100 magazines from 1983 to the present can be TYPEd, DISPLAYed or PRINTed using Format 9. To search the text to these articles, use MAGAZINE ASAP (File 647). Indexing is also included in NEWSEARCH (File 211) for daily updating during the current month.

MEDLINE (MED onLINE) (National Library of Medicine)

MEDLINE, produced by the U.S. National Library of Medicine (NLM), is one of the major sources for biomedical literature materials. MEDLINE corresponds to three printed indexes: Index Medicus, Index to Dental Literature, and International Nursing Index. Additional materials not published in Index Medicus are included in the MED-LINE database in the areas of communication disorders, and population and reproductive biology. MEDLINE is indexed using NLM's controlled vocabulary, MeSH (Medical Subject Headings). Abstracts, which are taken directly from the published articles, are included for over 47% of the records added from 1975 forward.

SCISEARCH (Institute for Scientific Information, Inc.)

SCISEARCH is a multidisciplinary index to the literature of science and technology prepared by the Institute of Scientific Information (ISI). It contains all the records published in Science Citation Index (SCI) and additional records from the Current Contents series of publications that are not included in the printed version of SCI. SCISEARCH is distinguished by two important and unique characteristics. First, journals indexed are carefully selected on the basis of several criteria, including citation analysis, resulting in the inclusion of 90 percent of the world's significant scientific and technical literature. Second, citation indexing is provided, which allows retrieval of newly published articles through the subject relationships established by an author's reference to prior articles.

SCISEARCH indexes all significant items (articles, review papers, meeting abstracts, letters, editorials, book reviews, correction notices, etc.) from approximately 4,500 major scientific and technical journals. Approximately 3,800 of these journals are further indexed by the references cited within each article, allowing for citation searching. The other 700 journals indexed have been drawn from the ISI publications: *Current Contents/Clinical Practice, Current Contents/Engineering, Technology & Applied Sciences,* and *Current Contents/Agriculture, Biology & Environmental Sciences.* SCISEARCH records contain one or more Research Fronts, which refer to specific highly active research areas.

BIBLIOGRAPHY

Ozone depletion. US gets tough on CFC emissions [news]
    Beardsley T
    Nature (ENGLAND)   Nov 13-19 1986,   324 (6093) p102, ISSN
    0028-0836

Carbon dioxide and its role in climate change.
    Benton GS
    Proc Natl Acad Sci U S A   Oct 1970,   67 (2) p898-9,   ISSN
    0027-8424

Seasonal variation of stroke--does it exist?
    Biller J; Jones MP; Bruno A; Adams HP Jr; Banwart K
    Department of Neurology, University of Iowa College of
    Medicine, Iowa City. Neuroepidemiology   1988,   7   (2)
    p89-98

[Environmental factors and asthma in the child population of
Vizcaya]
Factores ambientales y asma en la poblacion infantil de Vizcaya.
    Casas Vila C; Albisu Echeberria MV; Salazar Echeverri
    M; Ceballos Bizcarret A; Municio Martin MA; Pocheville
    Guruzeta I; Albisa   Echeberria J
    Departamento de Pediatria, Hospital de Cruces, Vizcaya.
    An Esp Pediatr (SPAIN)   May 1988,   28 (5) p401-4

Biological UV-doses and the effect of an ozone layer depletion.
    Dahlback A; Henriksen T; Larsen SH; Stamnes K
    Photochem Photobiol (ENGLAND)   May 1989,   49 (5) p621-5

Sunscreen oddity: success sharpens competition.
    Davis, Donald A.
    Drug & Cosmetic Industry VOL.: v145 ISSUE: n2 PAGINATION:
    p44(3) PUBLICATION DATE: August, 1989

'Very small' threat to public health; new study downplays ozone
depletion peril.
    Dembart, Lee
    Los Angeles Times VOL.: v103 SECTION: I PAGINATION: p27
    COLUMN   NUMBER: col 1 PUBLICATION DATE: Feb 25, 1984

Changes in the geographical distribution of malaria throughout
history.
    de Zulueta J
    Casa de Mondragon, Ronda, Malaga, Spain.
    Parassitologia (ITALY)   May-Dec   1987,   29   (2-3)
    p193-205

Clinical climatology.
    Diffey BL; Larko O
    Photodermatol   Feb 1984,   1   (1) p30-7,   ISSN 0108-9684

Poisons and precarious balance; one pollution may protect against another. (column)
> Easterbrook, Gregg
> Los Angeles Times   VOL.: v108 SECTION: V PAGINATION: p1
> COLUMN   NUMBER: col 1 PUBLICATION DATE: August 20, 1989

Health and the ozone layer [letter]
> Elwood JM
> BMJ (ENGLAND)   Dec 17 1988,   297 (6663) p1612-3,   ISSN 0267-0623

Health and the ozone layer [letter]
> Elwood JM
> BMJ (ENGLAND)   Oct 8 1988,   297 (6653) p918,   ISSN 0267-0623

Atmospheric ozone and man-made pollution
> Fabian P
> Naturwissenschaften   Jun 1976,   63 (6) p273-9,   ISSN 0028-1042

Skin cancer, melanoma, and sunlight.
> Fears TR; Scotto J; Schneiderman MA
> Am J Public Health   May 1976,   66 (5) p461-4,   ISSN 0090-0036

The problem with the climate
> DAS KLIMAPROBLEM. ANSATZE EINER POLITISCHEN LOSUNG
> Fischer W.; Di Primio J.C.; Sassin W.
> Forschungsgruppe   Wirtschaft,   Energie,   Investitionen in der Kernforschungsanlage Julich, Julich   Germany, Federal Republic of ENERGIEWIRTSCH. TAGESFRAGEN (Germany, Federal Republic of) , 1989,   39/5 (278-283)

Simulated stratospheric ozone depletion and increased ultraviolet radiation: effects on photocarcinogenesis in hairless mice.
> Forbes PD; Davies RE; Urbach F; Berger D; Cole C
> Cancer Res   Jul 1982,   42 (7) p2796-803,   ISSN 0008-5472

Climatic changes in the eyes of Eskimos, Lapps and Cheremisses.
> Forsius H
> Acta Ophthalmol (Copenh)   1972,   50 (4) p532-8

Letter: Stratospheric ozone destruction and halogenated anaesthetics.
> Fox JW; Fox EJ; Villanueva R
> Lancet   Apr 12 1975,   1 (7911) p864

Wavelength   dependence of pyrimidine dimer formation in DNA of
human skin irradiated in situ with ultraviolet light.
   Freeman SE; Hacham H; Gange RW; Maytum DJ; Sutherland JC;
   Sutherland BM   Biology Department, Brookhaven National
   Laboratory, Upton, NY 11973.
   Proc Natl Acad Sci U S A (UNITED STATES)   Jul 1989,   86
   (14)   p5605-9, ISSN 0027-8424

[The   possibility   of climate change due to the increasing CO_2
content in the atmosphere]
Uber   die Moglichkeit einer Klimaanderung durch den steigenden
CO_2-Gehalt der Atmosphare.
   Georgii HW
   Hippokrates   May   1978,   49   (2) p176-8,   ISSN 0018-2001

Broader aspects of clinical toxicology.
   Golberg L
   Clin Toxicol   May 1980, 16 (3) p365-70

[A t o p i c   n e u r o d e r m a t i t i s .   C l i n i c a l
course--pathophysiology--therapy]
Neurodermitis atopica. Klinik--Pathophysiologie--Therapie.
   Gloor M
   Fortschr Med   May 26 1983,   101 (20) p919-23,   ISSN
   0015-8178

Global stratospheric ozone and UVB radiation [letter]
   Grant WB
   Science (UNITED   STATES)   Nov   25   1988,   242 (4882)
   p1111-2

Skin cancer and the ozone layer [letter]
   Gray N
   Lancet (ENGLAND)   Jun 10 1989,   1 (8650) p1337,   ISSN
   0023-7507

Skin tests in Nigerian asthmatics from the equatorial forest
zone in Benin, Nigeria.
   Haddock DR; Onwuka SI
   Trans R Soc Trop Med Hyg   1977, 71 (1) p32-4

Contrast agents and the ozone layer [letter]
   Hall FM
   AJR Am J Roentgenol (UNITED STATES)   Sep 1989,   153 (3)
   p654-5, ISSN 0361-803X

Ultraviolet radiation at high latitudes and the risk of skin
cancer.
   Henriksen K; Stamnes K; Volden G; Falk ES
   Auroral Observatory, University of Tromso, Norway.
   Photodermatol (DENMARK)   Jun 1989,   6 (3) p110-7,   ISSN
   0108-9684

Solar radiation and malignant melanoma of the skin.
    Houghton AN; Viola MV
    J Am Acad Dermatol   Oct 1981,   5 (4) p477-83,   ISSN
    0190-9622

The changing climate of America.
    Hudson CL
    Nebr Med J (United States)   Jul 1967,   52 (7) p312-7

Air quality: unacceptable.
    Jaret, Peter
    Health   VOL.: v21 ISSUE: n3 PAGINATION: p48(4) PUBLICATION
    DATE: March,   1989

Ozone depletion and cancer risk.
    Jones RR
    St John's Hospital for Diseases of the Skin, London.
    Lancet (ENGLAND)   Aug 22 1987,   2 (8556) p443-6,   ISSN
    0023-7507

The stories we live by: personal myths guide daily life.
    Keen, Sam
    Psychology Today VOL.: v22 ISSUE: n12 PAGINATION: p42(6)
    PUBLICATION DATE: Dec, 1988

Ozone hole in stratosphere threatens life on the planet. (column)
    Konow, Al
    Whole Life Times ISSUE: n75 PAGINATION: p21(1)
    PUBLICATION DATE: June, 1988

Ozone depletion: implications for the veterinarian.
    Kopecky KE
    J Am Vet Med Assoc   Sep 15 1978,   173 (6) p729-33,   ISSN
    0003-1488

Impact of ozone depletion on skin cancers.
    Kripke ML
    Department of Immunology, University of Texas M.D.
    Anderson   Cancer Center, Houston 77030.
    J Dermatol Surg Oncol   Aug 1988,   14 (8) p853-7,   ISSN
    0148-0812

[Non-specific   triggering factors in 197 non-pollenogenic
asthmatics with an unfavorable outcome]
Factores desencadenantes inespecificos en 197 asmaticos, no
polinicos, de curso desfavorable.
    Lapena Lopez de Armentia S; Blanco Quiros A; Linares
    Lopez P;   Andion Dapena R; del Real Llorente M
    Departamento de Pediatria, Hospital Clinico, Valladolid.
    An Esp Pediatr (SPAIN)   Dec 1987,   27 (6) p441-4

A vision of health in the 21st century: medical response to the greenhouse effect.
    Last JM
    Department of Epidemiology and Community Medicine, University of Ottawa. Can Med Assoc J (CANADA) Jun 1 1989, 140 (11) p1277-9, ISSN 0008-4409

Solar considerations in the development of cutaneous melanoma.
    Loggie BW; Eddy JA
    Div. of Surgical Oncology, Cook County Hospital, Boulder, CO.
    Semin Oncol (UNITED STATES) Dec 1988, 15 (6) p494-9, ISSN 0093-7754

Cutaneous malignant melanoma and ultraviolet radiation: a review.
    Longstreth J
    ICF/Clement Associates, Inc., Fairfax, VA 22031.
    Cancer Metastasis Rev (UNITED STATES) Dec 1988, 7 (4) p321-33, ISSN 0891-9992

Planetary medicine. (Earth's environmental crisis)
    Lovelock, James
    American Health: Fitness of Body and Mind VOL.: v8 ISSUE: n2 PAGINATION: p86(3) PUBLICATION DATE: March, 1989

Health and the ozone layer [letter]
    Mackie RA; Rycroft MJ
    BMJ (ENGLAND) Nov 12 1988, 297 (6658) p1271-2, ISSN 0267-0623

Health and the ozone layer [editorial]
    Mackie RM; Rycroft MJ
    Br Med J [Clin Res] (ENGLAND) Aug 6 1988, 297 (6645) p369-70, ISSN 0267-0623

The changed pattern of malaria endemicity and transmission at Amani in the eastern Usambara mountains, north-eastern Tanzania.
    Matola YG; White GB; Magayuka SA
    J Trop Med Hyg Jun 1987, 90 (3) p127-34

Ozone depletion would have dire effects.
    Maugh TH 2d
    Science Jan 25 1980, 207 (4429) p394-5, ISSN 0036-8075

A blueprint for the environment.
    McKee, Steve
    American Health: Fitness of Body and Mind VOL.: v8 ISSUE: n2 PAGINATION: p88(2) PUBLICATION DATE: March, 1989

Biological   amplification   factor   for   sunlight-induced
nonmelanoma skin cancer at high latitudes.
    Moan J; Dahlback A; Henriksen T; Magnus K
    Institute for Cancer Research, Montebello, Oslo, Norway.
    Cancer   Res (UNITED   STATES)   Sep 15 1989,   49   (18)
    p5207-12,   ISSN 0008-5472

Ozone   depletion   and   its   consequences   for the fluence of
carcinogenic sunlight.
    Moan J; Dahlback A; Larsen S; Henriksen T; Stamnes K
    Institute for Cancer Research, Norwegian Radium Hospital,
    Oslo,   Norway.   Cancer   Res (UNITED   STATES)   Aug 1
    1989,   49   (15)   p4247-50,   ISSN 0008-5472

[Ozone depletion, ultraviolet rays and skin cancer]
Ozonlag, ultrafiolett str.ANG.aling og hudkreft.
    Moan J; Larsen S; Dahlback A; Henriksen T
    Tidsskr   Nor Laegeforen (NORWAY)   Nov 10 1988,   108 (31)
    p2838-40,ISSN   0029-2001

Halothane anaesthetic and the ozone layer [letter]
    Norreslet J; Friberg S; Nielsen TM; Romer U
    Lancet (ENGLAND)   Apr 1 1989,   1 (8640) p719,   ISSN
    0023-7507

Increased   outdoor   recreation,   diminished   ozone layer pose
ultraviolet radiation threat to eye [news]
    Olson CM
    JAMA (UNITED STATES)   Feb 24 1989,   261 (8) p1102-3,   ISSN
    0098-7484

Saving our ozone layer: the DuPont Company's decision [editorial]
    Paulshock BZ
    Del Med J (UNITED STATES)   Sep 1988,   60 (9) p533-4,   ISSN
    0011-7781

All   tanning   is   called 'toxic injury.' (by National Institutes
of Health panel)
    Pollner, Fran
    Medical World News VOL.: v30 ISSUE: n11 PAGINATION: p15(2)
    PUBLICATION DATE: June 12, 1989

Seasonal   change   and   its   effect   on   the prevalence of
infectious skin disease in a Gambian village.
    Porter MJ
    Trans R Soc Trop Med Hyg   1980,   74 (2) p162-8

Stratospheric   ozone   depletion.   A   proposed   solution to the
problem.
    Redman JC
    University of New Mexico School of Medicine, Albuquerque.
    Am J Dermatopathol   Oct 1987,   9 (5) p457-8,   ISSN
    0193-1091

Actinic DNA damage and the pathogenesis of cutaneous malignant
melanoma.
    Ross PM; Carter DM
    Laboratory for Investigative Dermatology, Rockefeller
    University, New York, New York.
    J Invest Dermatol (UNITED STATES)   May 1989,   92 (5 Suppl)
    p293S-296S,

Atmospheric CO2 consequences of heavy dependence on coal.
    Rotty RM
    Environ Health Perspect   Dec 1979,   33 p273-83,   ISSN
    0091-6765

Projections of increased non-melanoma skin cancer incidence due
to ozone depletion.
    Rundel RD; Nachtwey DS
    Photochem Photobiol   Nov 1983,   38 (5) p577-91,   ISSN
    0031-8655

[Effect   of   the   atmospheric   ozone   layer   on   the
biologically active ultraviolet radiation on the earth's surface]
Der   Einfluss   der Ozonschicht der Atmosphare auf die biologisch
wirksame Ultraviolettstrahlung an der Erdoberflache
    Schulze R; Kasten F
    Strahlentherapie   Aug 1975,   150 (2) p219-26,   ISSN
    0039-2073

Biologically   effective   ultraviolet   radiation: surface
measurements in the United States, 1974 to 1985.
    Scotto J; Cotton G; Urbach F; Berger D; Fears T
    Biostatistics Branch, National Cancer Institute, Bethesda,
    MD   20892.
    Science (UNITED   STATES)   Feb   12   1988,   239   (4841 Pt 1)
    p762-4,   ISSN 0036-8075

Skin cancer epidemiology: research needs.
    Scotto J; Fears TR
    Natl Cancer Inst Monogr   Dec 1978,   (50) p169-77,   ISSN
    0083-1921

Nitrous oxide and the greenhouse effect [letter]
    Sherman SJ; Cullen BF
    Anesthesiology (UNITED   STATES)   May   1988,   68   (5)
    p816-7, ISSN 0003-3022

Ozone depletion & your health. (column)
    Sibbison, Jim
    Bestways   VOL.: v16 ISSUE: n1 PAGINATION: p60(1)
    PUBLICATION DATE: Jan, 1988

Respiratory syncytial virus infection in north-east England.
    Sims DG; Downham MA; McQuillin J; Gardner PS
    Br Med J   Nov 6 1976,   2 (6044) p1095-8

Non relationship of climatologic factors and painful sickle cell anemia crisis.
　　Slovis CM; Talley JD; Pitts RB
　　J Chronic Dis　1986,　39 (2) p121-6

Spectral dependencies of killing, mutation, and transformation in mammalian cells and their relevance to hazards caused by solar ultraviolet radiation.
　　Suzuki F; Han A; Lankas GR; Utsumi H; Elkind MM
　　Cancer Res　Dec 1981,　41 (12 Pt 1) p4916-24

Dilemmas in the development of a greenhouse policy
　　Swart R.J.; De Boois H.
　　Rijksinstituut voor Volksgezondheid en Milieuhygiene (RIVM), Bilthoven Netherlands
　　MILIEU (Netherlands)　, 1989, 4/3 (73-78)

'Healthy tan' - a fast-fading myth.
　　Sweet, Cheryl A.
　　FDA Consumer　VOL.: v23 ISSUE: n5 PAGINATION: p11(3)
　　PUBLICATION DATE: June, 1989

Are sunscreens a skin cancer smokescreen?
　　Thomson, Bill
　　East West　VOL.: v19 ISSUE: n1 PAGINATION: p40(3)
　　PUBLICATION DATE: Jan, 1989

Conservation of climate is common responsibility throughout the world.　Tickell, Crispin
　　Oil Daily ISSUE: n9149 PAGINATION: pB12(1)
　　PUBLICATION DATE: Nov 14, 1988

Causes and effects of stratospheric ozone reduction: an update.
　　Urbach F
　　J Am Acad Dermatol　Aug 1982,　7 (2) p271-3

Hot-house effect induces cultural revolution
　　Van der Jagt C.
　　Netherlands
　　MILIEU DEF. (Netherlands)　, 1989, 18/2 (7-10)

Ozone depletion and skin cancer.
　　van der Leun JC
　　University of Utrecht, Institute of Dermatology, The Netherlands.
　　J Photochem Photobiol -B- (SWITZERLAND)　May 1988,　1 (4) p493-4,　ISSN 1011-1344

Solar cycles and malignant melanoma.
　　Viola MV; Houghton A; Munster EW
　　Med Hypotheses　Jan 1979,　5 (1) p153-60,　ISSN 0306-9877

Solar radiation and cutaneous melanoma.
    Viola MV; Houghton AN
    Hosp Pract [Off]   Sep 1982,   17 (9) p97-106,   ISSN
    8750-2836

[Solar activity, dynamics of the ozone layer and possible
role of ultraviolet radiation in heliobiology]
Solnechnaia   aktivnost',   dinamika   ozonosfery   i
vozmozhnaia rol' ul'trafioletovogo izlucheniia v geliobiologii.
    Vladimirskii BM
    Kosm Biol Aviakosm Med   Jan-Feb 1982,   16 (1) p12-5,   ISSN
    0302-5969

Effect   of   varying dose of UV radiation on mammalian skin:
simulation of decreasing stratospheric ozone.
    Willis I; Menter JM
    J Invest Dermatol   May 1983,   80 (5) p416-9

National   environment   of   early   food   production   north of
Mesopotamia. Climatic change 11,000 years ago may have set
the stage for primitive farming in the Zagros mountains.
    Wright HE Jr
    Science (United States)   Jul 26 1968,   161 (839) p334-9

The   evaluation of immediate hypersensitivity reactions: current
concepts and future directions.
    Yoshida S; Halpern G; Gershwin ME
    Department of Internal Medicine, University of California,
    Davis.
    Allergol Immunopathol (Madr) (SPAIN)   Nov-Dec 1987,   15 (6)
    p335-41

Study warns of greenhouse effect [news]
    J Am Vet Med Assoc (UNITED STATES)   May 1 1989,   194 (9)
    p1175, ISSN 0003-1488

Anaesthetic agents and the ozone layer [letter]
    Lancet (ENGLAND)   May 27 1989,   1 (8648) p1209-10,   ISSN
    0023-7507

Greenhouse effect [letter]
    Lancet (ENGLAND)   May 27 1989,   1 (8648) p1208-9,   ISSN
    0023-7507

Anaesthetic agents and the ozone layer [letter]
    Lancet (ENGLAND)   May 6 1989,   1 (8645) p1011-2,   ISSN
    0023-7507

Ozone depletion update.
    Medical SelfCare ISSUE: n48 PAGINATION: p13(1)
    PUBLICATION DATE: Sept-Oct, 1988

Health in the greenhouse. (atmospheric greenhouse effect)
(editorial)
     Lancet   VOL.: v1 ISSUE: n8642 PAGINATION: p819(2)
     PUBLICATION DATE: April 15, 1989

INFORMATION NEEDS AND RESEARCH PRIORITIES

Michael P. Farrell, Paul Kanciruk, and
Frederick M. O'Hara, Jr.[1]

Environmental Sciences Division
and
The Center for Global Environmental Studies
Oak Ridge National Laboratory
Oak Ridge, TN 37830

Introduction

The U.S. Global Research Plan (Committee on Earth Sciences, 1989) and the International Geosphere-Biosphere Programme (IGBP, 1988) were created to assess the effects of global climate change but have not been able to devote much attention to the consequences climate change will have on human health and welfare. Although researchers and policy makers recognize that climate change will have complex effects on resources (Fig. 1), in general, the social and medical sciences have not received appropriate national attention under the banner of global change. To address this imbalance, the public health research community needs to launch a national coordinated effort so that the social and medical sciences are as fully represented as other scientific disciplines. At this time, the social sciences have gained a foothold primarily through policy and socioeconomic analyses, yet little attention is paid to public health. Indeed, funding for public health research is low even in comparison with support for other social sciences.

The scientific community's low interest in and lack of acceptance of public health studies is unfortunate because public health issues are a major challenge for global climatology. In 1985 the U.S. Department of Energy (DOE) produced a series of reports on global climate change. One characterized the additional information required to understand and deal with climate change. The report covered various resources, such as fisheries, water, forest and agriculture. In addition, it included a solitary chapter on

---

1   Private Consultant

© 1990 by Elsevier Science Publishing Co., Inc.
Global Atmospheric Change and Public Health
James C. White, Editor

human health, which was extremely difficult to develop (White and Hertz-Picciotto, 1985). At that early stage in the investigation of global climate change (about 1983), the challenges posed by that chapter demonstrated how difficult it is to address public health issues in a global context.

The difficulty arises from three deficiencies in our understanding and advancement of public health science: lack of recognition of and consensus on the relevant questions and research directions that should be pursued, lack of comprehensive models of health-related processes, and lack of world-wide data on health problems. When these shortcomings are successfully addressed, public health science will be able to measure the human dimensions of global climate change.

Needed Analyses

White and Hertz-Picciotto (1985) identified a number of questions about the effects on human health of changing climate:

—   Would a warmer climate affect human thermoregulation, acclimation or adaptation?

—   Would it produce any changes in human physiology or biochemistry?

—   Would it increase or decrease the incidence of birth defects?

—   Would heat waves, storms, lightning and floods take higher tolls?

—   Would more extreme climates increase world-wide morbidity and mortality?

—   Would warmer, moister climates enhance the breeding and spread of bacteria, viruses, fungi or pollen?

—   Would more turbulent air movement and more extreme temperatures and humidity spread airborne diseases more effectively?

—   Would diseases carried by humans become more prevalent and widespread?

—   Would vector-borne viruses, rickettsia and bacteria be spread more readily?

—   Would parasitic diseases thrive better in a warmer climate?

—   Would climate affect agricultural production, the availability of food for humans, human nutrition, human resistance to disease, and human health?

—   What interactions between disease and nutrition might occur?

—   Might air pollution be affected by changes in temperature, in precipitation, in wind speed, and in land cover?

— Would climate change affect the quality and quantity of outdoor recreation the human population might enjoy and, if so, with what results in terms of health?

Needed Models

Specific answers to these questions are not immediately forthcoming. Indeed, we do not yet know all the questions that should be asked about the health-related impacts of global climate change. Public health researchers need more sophisticated tools to perform these analyses with reliability. Researchers currently use statistical or simulation models and small-scale information (information derived from a small sampling of a large population or from an intensive sampling of a geographically small area) to produce an analysis of public health risk on a global scale.

To carry out such predictive analyses successfully, researchers need new simulation models that can account for relationships between or among entities or processes. Such quantitative models would be very unlike the conceptual models that public health professionals have used so successfully to eliminate smallpox and to control polio. Unfortunately, the current conceptual models are not robust and cannot deal with problems like global climate change because a large number of variables and large uncertainties are involved.

As a result of their lack of robustness, the health sciences' conceptual models and their results have not been fully accepted by the scientific and policy-making communities that deal with global change. To gain acceptance, the health sciences need to develop a quantitative approach that can address uncertainties and successfully model the processes involved. To develop such a model, however, researchers must first thoroughly analyze the processes. Once the processes are analyzed and modeled, regional statistics can be entered into the model. Then public health researchers can perform quantitatively-based global analyses.

Needed Data

The lack of depth in the available data will severely hamper public health research into global climate change. Public health researchers frequently do not have enough information available to understand current conditions, let alone predict the future. Other areas of scientific inquiry into the causes and probable effects of global climate change are much more mature than public health studies. For example, not only do thousands upon thousands of data points and hundreds of journal articles describe the chemistry of the carbon cycle, but many reviews of the global carbon cycle tell us exactly what we do not know, why we do not know it, and what resources are needed to develop a fuller understanding. Such self-review is vital to the development of science. Thus, similar analyses are needed for public health issues.

Public health officials and researchers should not despair, however; similar shortcomings have existed in almost every other resource area addressed to assess the secondary impacts of climate change. There is hope. On-line bibliographic data bases, directories of numeric data bases, and referral services like the National Environmental Data Referral Service are certainly helpful and often necessary. Yet, in order to assess what the real problems are, researchers need to assemble and compile masses of raw data. For a while, analysis within national boundaries will probably suffice, and the work will entail expanding such studies as the St. Louis Study or performing studies of regional corridors. Soon, however, national studies and global analyses (city-by-city and region-by-region) will have to be performed.

Such efforts will produce a tremendous amount of data, all of which will need to be collected, stored, uniformly formatted, assessed for quality, made available to the scientific community, and distributed to researchers. Tasks like these have historically been carried out by centralized data centers funded by the government, academic institutions, professional organizations, and private corporations. As for global change issues, we should recognize that analyzing global-scale issues will require data bases that are much larger than those currently produced. Also, because the data collected must come from a multiplicity of disciplines, public health data will need to be correlated with data on climate, agricultural production, populations and water resources, among many other types of data. Data, or the lack of it, will ultimately be a potential problem to be addressed by public health researchers.

The amount and diversity (in both subject matter and source) of data that will be produced in assessing global problems will overwhelm a traditional data center and outstrip its capabilities. The study of global public health issues will require a data management system that can provide not only the vehicle for exchanging extant, acquired information but also the mechanisms for adding value to existing information, synthesizing new data sets, and initiating networking within the research community. What will be needed is an Information Analysis Center (IAC) (Fig. 2).

The IAC as a Tool

An IAC would not be just a place to store and distribute data. Such a center would apply innovative approaches to managing research data, thereby increasing the number of information resources and producing enhanced data and information products to support a broadly defined user community. Justifiable under the purview of an IAC would be such activities as compiling bibliographies, conducting user-defined customized bibliographic searches, publishing newsletters, sponsoring workshops, and producing directories of researchers and policy makers. Its functions might include (but not be limited to) the following:

- Acting as the central data hub for a research program, providing program management with a clearinghouse for sorting available information and data, and serving as the conduit for both the distribution of data generated by the program and the acquisition of external data needed by the program.

- Setting standards not only in the format and transfer of information but also in the overall quality and usefulness of data to be included in the center, the depth of documentation, the extent and type of quality assurance needed, the type and breadth of data distribution, and the methods of archiving.

- Identifying what data are most needed by the general research community and deciding which data should be obtained, processed, distributed and archived.

- Taking an active interest in providing data to address issues at hand by assembling information from multiple sources or by producing larger, more numerous, and higher-level data products that are not of the type normally produced by researchers or policy makers.

- Providing quality assurance that greatly increases confidence in reliability: for data sets, such quality assurance includes checking completeness, identifying unreasonable values or inconsistent correlations, and culling questionable observations; for computer models it includes analyzing the computer codes for errors and sensitivity. All quality assurance would be accomplished in close consultation with, and approval by, the original data supplier.

- Documenting data so that someone not familiar with the data could fully understand and use the data 20 years from now solely on the basis of the documentation.

- Widely notifying potential users that data are available, copying and sending data to researchers in a usable form, and providing input-output routines for that medium, as well as including summary statistics with the data.

- Archiving data in a way that ensures its integrity and usability as storage technology advances and as older equipment and media become obsolete.

- Promoting interaction between individual researchers, organizations, data centers, and other IACs, ensuring that any entity can directly receive information from or provide information to any other entity through newsletters, conferences, workshops, direct contacts, participation in bilateral and multilateral agreements, and physical and electronic mail exchange of requests and data (Fig. 3).

Thus an IAC links many communication activities within the scientific community and serves many of the information needs of researchers (Fig. 4).

The CDIAC Model

The Carbon Dioxide Information and Analysis Center (CDIAC) has served the international research community since 1982 under the guidance of the U.S. Department of Energy. CDIAC covers a very small part of the scientific spectrum: the three general areas of climate and weather records, the carbon cycle (including both oceanic and vegetative components), and vegetation. Currently, CDIAC distributes 35 data bases to businesses, universities and government agencies. Figure 5 shows how CDIAC shares and contributes to the research process.

Our most important product is data bases that have undergone a rigorous quality assurance process. We review the data base, cull questionable data points, reformat it for distribution and use, document it, and perhaps use it in demonstration analyses. However, the data base is not released to the public until the principal investigator signs an approval expressing agreement with the final product (Fig. 6). Indeed, we do not adopt a data base unless we have the full cooperation of the principal investigator, even though the data base might be very critical. Such authority over the data is very important because the data belong to and reflect on the person who produced the information.

The data base is then 'beta tested.' In this process, the data base is sent to a research site, where specialists work intensively with it for a few months to discover any inconsistencies. Once we are assured that the data are reliable, the data base is permanently archived so that it can be requested and obtained by any researcher. As an additional quality check, we survey the people who request and use the data, asking them, "Did you find any problems? What can we do to improve the data base?" and so on. If any error is discovered in the data base, we notify the user community, or we correct the error and announce that a new version of the data base is available.

Quality Assurance, A Special Concern

With data bases of this size and from this diversity of sources, the quality of the data is bound to vary. As a result, the problem of quality assurance has to be given special attention. Variations in data quality arise from two major shortcomings in the way science is conducted.

The first shortcoming is that researchers generally are not rewarded for producing high-quality data bases in the way they are rewarded for high-quality publications. Indeed, compiling data bases takes time and money away from research. National research programs — such as space, health and education — have been supportive of data base compilation largely in times of plentiful funding. When money has become scarce, data management has frequently been cut first. Fortunately, such is not always the case. The global change program is an excellent example of a program where, at least in the planning stage, data collectors have become equal partners with scientists.

The second shortcoming is that researchers are not particularly good at assuring the quality of their data; they are slightly myopic. Researchers might, for example, overlook spurious data or outliers. Gaining some distance from the work gives new insight into, and a less impassioned perspective towards, inconsistencies in the data. Quality assurance checks by independent reviewers often find problems in the data that researchers have overlooked.

As a result, IACs can expend a significant amount of their resources on quality assurance. For example, when CDIAC worked with the United Nations to assemble a data base on the global amounts of CO_2 emissions from fossil-fuel burning, quality assurance and upgrading the data base were not trivial. Developing an accurate data base might be a multimillion dollar investment. Yet the amount spent on assuring the data's quality is justified, especially when one considers how much research is represented by that data base and how much could be depending on its analysis.

To gain an appreciation of the amount of culling necessary to ensure high quality in a data base, consider the Historical Climatology Network. CDIAC started with data from ~9000 measuring stations in the United States and applied very rigid quality assurance procedures to that data base. Researchers dismissed some climate data records because they did not cover a long enough period, dismissed others because the data-collecting stations were moved, and dismissed yet others for other inconsistencies in the data. Much of this quality control was provided by the National Climate Data Center in Asheville, North Carolina. CDIAC, as the technical liaison, produced the final documentation. At the end of the quality control process, data from only 1200 stations remained for global climate change analysis. Because calculations and predictions made thorough use of upgraded data bases, they will have much greater meaning, reliability and authority.

An IAC for Public Health

To establish an IAC capability for public health researchers, funds must be solicited and secured. Because of the massive amount of data to be handled, the sheer number of investigators conducting the research, the variety of scientific disciplines represented, and the geographic dispersal of the researchers, no single site will be able to do all the work of the IAC. The work load will have to be borne cooperatively by a variety of organizations, and services will have to be paid for through subcontracts or grants from institutions. It would be reasonable to put sums of money in various sites and to have each site handle a particular aspect of the problem.

To ensure that the approach to the problem remains unified, certain aspects of the program will have to be coordinated centrally, perhaps by the federal government. The provision of centralized program office support is essential because of the scope and range of the enterprise: at CDIAC, we have identified 5000 people in 150 countries who are involved in research on global climate change. The needs of this diverse community cannot be met successfully through the uncoordinated efforts of scores of institutions scattered around the world.

The research community that can contribute to a fuller understanding of public health issues must be identified. Then information can be addressed to that target group through the educational milieu. For example, at CDIAC, we continuously target 5000 global change researchers, asking them for additional research results and giving them the latest available information.

An IAC in public health research will have a variety of other benefits. Such an IAC will promote networking (as mentioned at this conference) which is a critical need in the public health research community. Networking will allow thousands of researchers from different institutions, agencies and countries to exchange comments and have input to establishing research priorities. In addition, an administrative program would determine the consensus of the research community about where the research should be directed and who should perform specific tasks. Such a dispersed organization coordinated by a central office will allow the field of public health to formalize the management of data collection, formatting, quality assurance, storage and use.

Some Strategic Considerations

Set Priorities and Narrow the Focus

Some global problems are long-term propositions (e.g. the accumulation of CO_2 in the atmosphere from the burning of fossil fuels), whereas other problems are relatively acute emergencies (e.g. the destruction of rain forests). Discriminating between long-term and near-term concerns helps researchers define priorities and sharpen the focus of their efforts. Such a distinction is embodied in an analysis recently completed for the U.S. Agency for International Development. In that report, ORNL scientists looked at a list of resources (e.g. fisheries, vegetable crops, agricultural animals and forests) and considered the effects that global climate change might have on those resources. The effects reflect both long-term and near-term changes (Fig. 7).

Similarly, in discussing public health effects, it might be strategically wise to distinguish between long-term effects and near-term emergencies and consider them separately. Although the public (and their elected representatives) can always doubt and temporarily ignore the long-term effects of climate change, they cannot argue about emergencies. Society must immediately respond to floods, disease and famine.

Retain a Global Perspective

Human populations cannot be the sole focus in global analyses. As researchers, we must consider the highly interdependent web of resources which is subject to a vast number of interactions. Indeed, some interactions only indirectly affect public health (e.g. the interactions between water resources, agriculture, nutrition and health) (Fig. 1). Although human populations certainly are important, the other interrelationships have not been accorded an appropriate share of the necessary resources; this imbalance must change.

Make Global Models Regionally Specific

Regional specificity will be critical to public health. Once the new climate regime in a particular small region is known, the public health effects of that climate will need to be predicted. Global estimates will not predict what will happen to individuals in specific locales. Therefore, the models of global processes must be able to predict changes in public health for regions of meaningful size and scale.

Conclusion

Research in the social and medical sciences has been sorely neglected in the debate about global climate change. In that debate, much has been said about the environment whereas little has been said about people. Such an oversight is a serious deficiency in our international planning efforts because the effects on people are ultimately what will be important. To rectify the oversight, social scientists as well as medical scientists must become more active in the debate.

Although the public health research community may be tempted to devote increased time, money and manpower to develop more research projects, the identification and pursuit of specific research projects should not be their sole focus at this time. Public health researchers must also develop a stringent philosophy of data and data management. Under that philosophy, data management and data quality assurance must be equal partners with research. Only then can public health researchers and other scientists fully appreciate each others' potential to contribute to the understanding of both global change and the effects that global change may have on our planet.

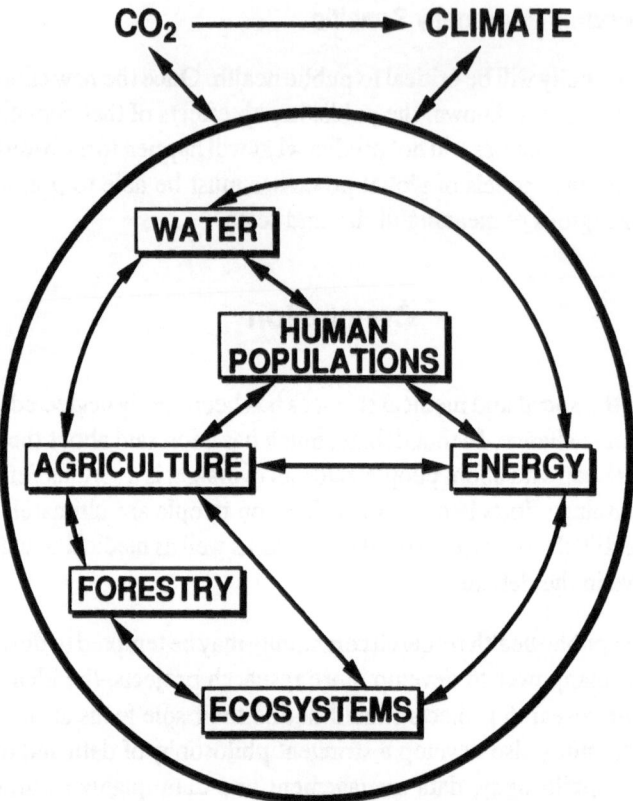

Figure 1.   Resource interactions important in global change. Climate change will
             have complex effects on natural resources and the human population.

The Information Analysis Center
Encompasses Many Functions

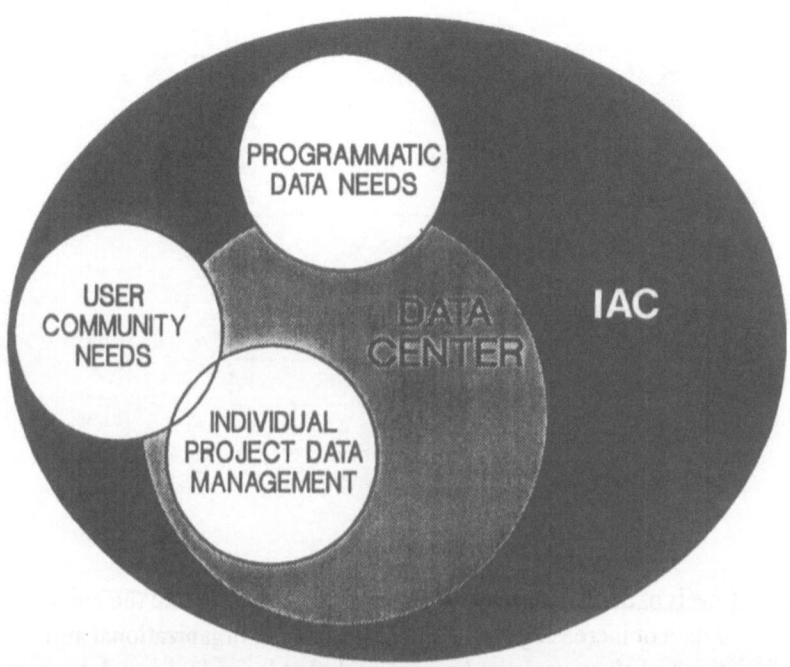

Figure 2.   Data management and user communities. The individual researcher undertaking data management can serve only a small portion of the user community and can rarely serve programmatic information needs. The data center expands this role but falls short of serving the entire user community and the majority of program needs. The information analysis center (IAC) serves the individual researcher, the larger user community, and program-level data needs through its role in data management, its variety of functions, and its value added concept of data management.

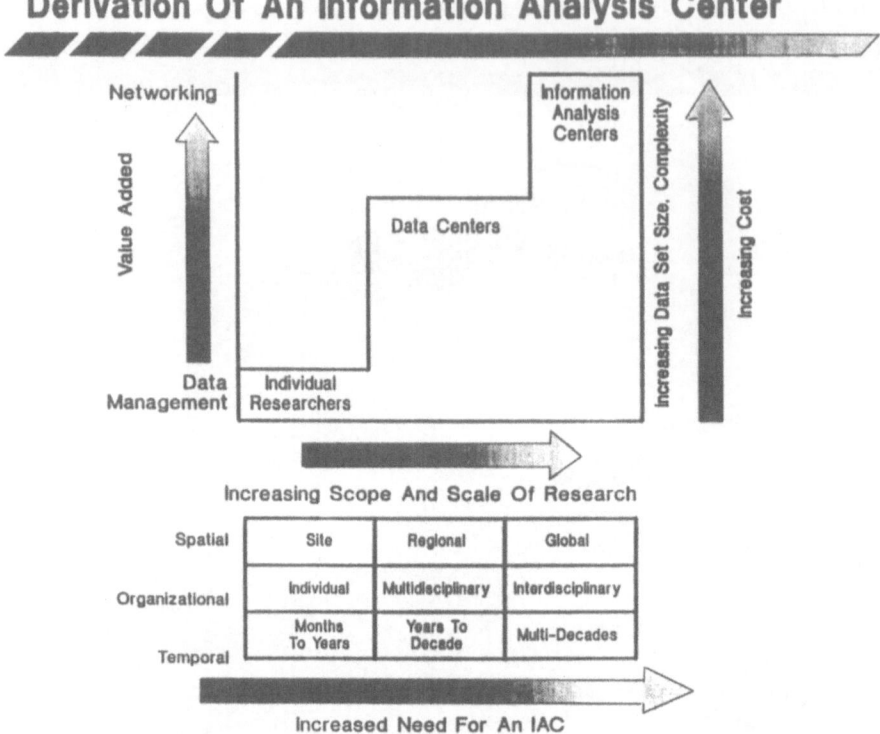

Figure 3.   Levels of data management. As research needs dictate the analysis of data of increasing scope and scale, spatial, organizational and temporal characteristics become such that individual and data center approaches to data management become inadequate. At the level of data management needed to address global issues and human health, the concept of an information analysis center (IAC) becomes justified.

The IAC Is A Gateway Linking Many Activities And Serving Many Needs

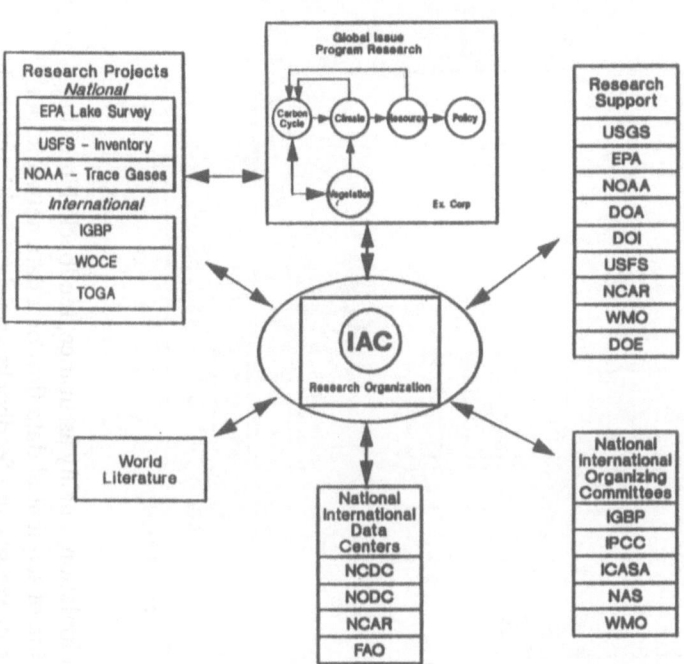

Figure 4.   The information analysis center (IAC). The IAC serves many functions, not the least of which is to act as a hub linking the research community with support and funding agencies and providing a focus for data management at the program level.

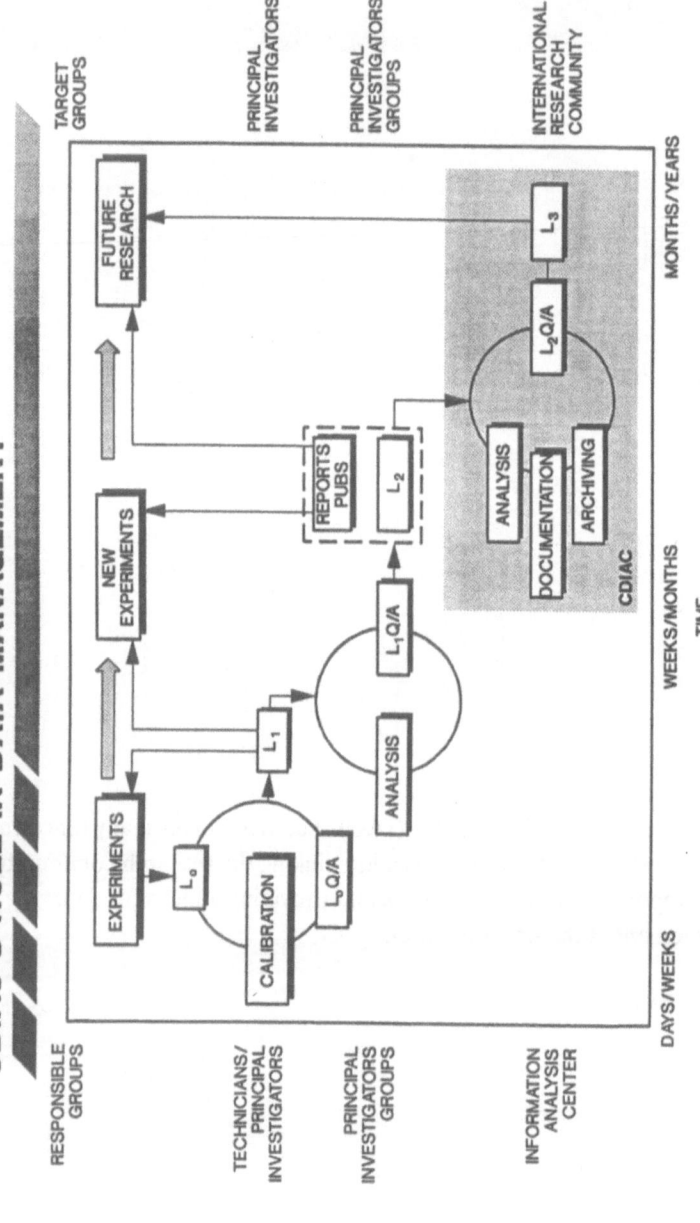

Figure 5. Levels of data manipulation, quality assurance, feedback. Note that this is not a depiction of sequential data flow or a data network, but a conception of the processes and feedbacks associated with research data. CDIAC's scope is shown in the shaded area.

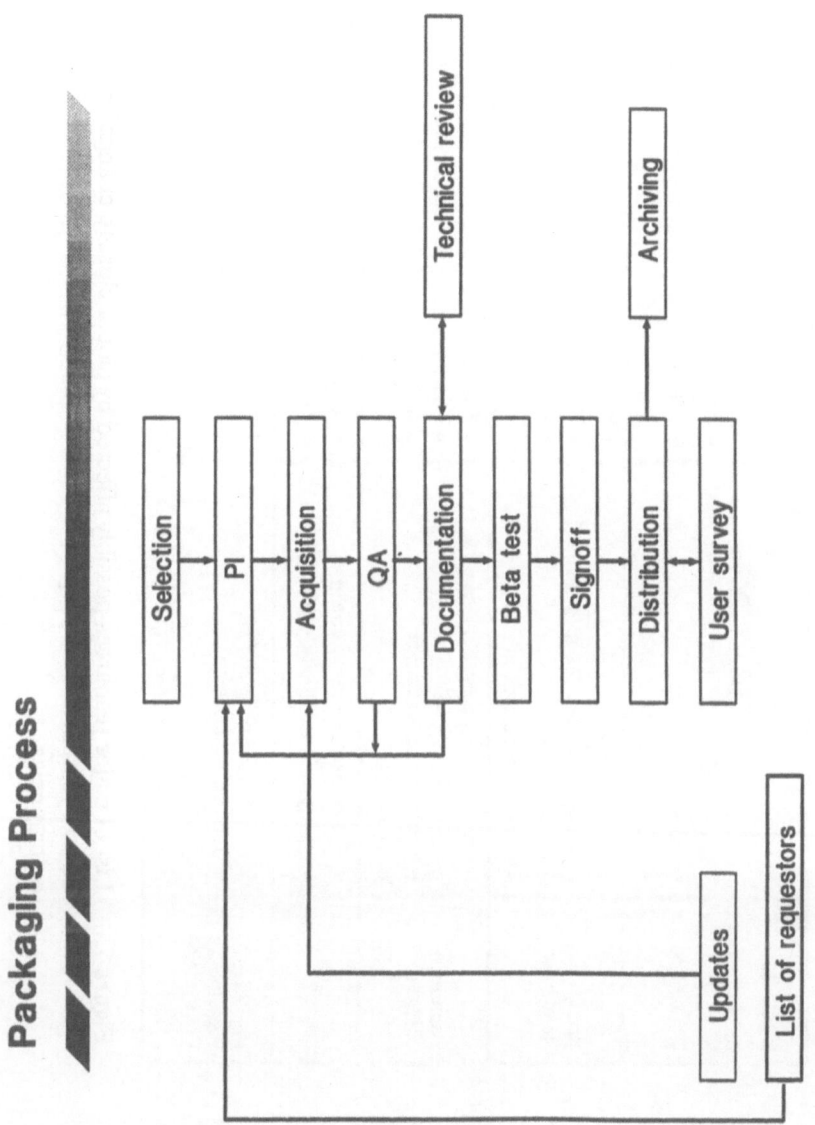

Figure 6.   Steps in the CDIAC numeric data packaging process.

Impacts On Resources

Resource	Impacts	
	Long-Term Change	Emergencies
Agriculture	• Crop productivity (increase/decrease) • Salinization • Locations of crop regions	• Increased crop failures and losses • Famine
Water Resources	• Salinization • Management and timing of run-off • Water supply • Flood management	• Drought • Flood • Salinization
Population & Health	• Infectious and parasitic diseases (incidence, range) • Nutrition and sanitation • Air pollution	• Epidemics • Respiratory and cardiac stress • Refugees
Energy	• Hydropower resource changes • Changes in electricity demand	• Supply system failure (e.g. loss of hydropower services, storm damage to supply system)
Forestry, Fisheries, Ecosystems	• Shift in ecological zones • Species composition • Salinization • Productivity change	• Fires • Pests and diseases • Flooding

Figure 7.   List of major resources possibly affected by global climate change broken down by impacts from long-term climate change and shorter-term emergencies.

Acknowledgment

This publication is based on research supported by the Carbon Dioxide Research Program, Atmospheric and Climate Research Division, Office of Health and Environmental Research, U.S. Department of Energy, under contract DE-AC05-84OR21400 with Martin Marietta Energy Systems, Inc. Publication number 3521 of the Environmental Sciences Division, Oak Ridge National Laboratory.

References

Committee on Earth Sciences, 1989. *Our Changing Planet: The FY 1990 Research Plan*. Committee on Earth Sciences, Office of Science and Technology Policy, Federal Coordinating Council on Science, Engineering, and Technology.

International Geosphere-Biosphere Programme, 1988. *Toward an Understanding of Global Change*. Committee on Global Change, U.S. National Committee for the IGBP, National Academy Press, Wash. D.C., 213 pp.

White, M.R. and I. Hertz-Picciotto, 1985. "Human Health: Analysis of Climate Related to Health." In: White, M.R. (ed.), *Characterization of Information Requirements for Studies of CO$_2$ Effects: Water Resources, Agriculture, Fisheries, Forests and Human Health*. U.S. Department of Energy, Office of Basic Energy Science, Wash.,D.C. DOE/ER-0236, 235 pp.

PANEL DISCUSSION: RESEARCH PRIORITIES

Moderator:

Gordon MacDonald, Vice President and Chief Scientist, The MITRE
Corporation

Panelists:

Janice Longstreth, Vice President, Division of Science Policy, Clement
Associates
Alexander Leaf, Chairman, Department of Preventive Medicine,
Harvard Medical School
Edwin M. Kilbourne, Chief, Health Studies Branch, Centers for Disease
Control
Suzanne Holmes Giannini, Center for Vaccine Development, University
of Maryland School of Medicine
Robert E. Shope, Director, Arbovirus Research Unit, Division of Infectious
Disease Epidemiology, Yale University School of Medicine
Michael P. Farrell, Director, CO_2 Information Analysis and Research
Program, Oak Ridge National Laboratory

DR. MACDONALD: In participating in the planning of this activity and looking at
a good deal of the literature on the subject and, in particular, examining the coverage in
the semipopular and even popular press, I've been impressed by the fact that the issue
of public health and climate change gets relatively little attention. If you look at the question of ozone depletion, clearly issues of incidence of skin cancer were a driving consideration back in the 1970s when that issue first surfaced and received public attention.
As much more has become known about ozone depletion of the stratosphere, that interest has been maintained. To a very large extent, public health has been equated with skin
cancer and not the wide range of other potential public health consequences. As far as
global warming is concerned, I would say that public health has received very little attention. Why is this so? If you look at recent cover stories in news magazines, you'll see
a lot of attention to sea level rise and to direct impacts on agriculture, yet not to the
secondary impacts, the physical effects. There is very little, if any, mention of impacts on
health.

DR. LONGSTRETH: Let me tell you about my experience in helping to write EPA's
report to Congress on global climate change. I was asked to write the chapter on health,
and it went to a scientific advisory board of the EPA for review. In the process of the
review, it was pointed out to me, gently but fairly firmly, that the only reason there was
a chapter on health in this particular document was because Congress had mandated

Published 1990 by Elsevier Science Publishing Co., Inc.
Global Atmospheric Change and Public Health
James C. White, Editor

235

that it be so. Clearly the effects on the other aspects of our world, that is agriculture and forestry and so forth, were going to far overwhelm anything that we would expect to see on health in the United States. Admittedly the report to Congress did focus only on the U.S., and I think that's the common world-wide perception. I must confess I was delighted to hear Dr. Farrell say that he thinks public health is where we need the most research because, as a health researcher, what I've been hit over the head with time and time again is, "The public health aspects are interesting and they're politically sensitive, but they're not going to be a big deal."

DR. LEAF: I think that the problem has been simply that, while the atmospheric scientists have been looking at these issues and collecting the data as well as they can, there's still so much controversy and so much uncertainty in the physical changes that are going to impact upon the globe. The public health community and the medical profession haven't felt that there was a credible body of information on which they could depend. I think those of you who heard my presentation realized that I was talking about a second-order degree of speculation. Is there still all the speculation among the physical scientists as to what the magnitude of the changes will be? You can only begin to talk about what the effects will be on nutrition and food supplies. It isn't surprising to me that the public health field has delayed its entry into this debate until there was a little better information to base speculations upon. But what Dr. Farrell was telling you is also correct in that there tends to be a compartmentalization of information in our world today. The people who are working in the field from the point of view of global changes think that it's their problem and they're not particularly interested in bringing in others to whom the information may also be relevant. I know that I've been trying to get the National Academy of Sciences to at least let some of us who are interested in the health effects participate in the expert committees or at least listen in on the expert committees, and they haven't even thought of the idea that the health community might find that the information was relevant.

With those kinds of perceptual gaps, it seems to me that, with the uncertainty in the physical science approach, it's not surprising that the conservative medical profession and public health scientists have said, "We have enough to work on and we're not going to jump into something that's so uncertain." The importance of a meeting like this is to begin to show the health professions that these issues are very big and loom very large in respect to the future health of our society and, therefore, it now is time for them to get involved.

DR. KILBOURNE: I fully agree with Dr. Leaf in terms of the impact of uncertainty on people's willingness to pursue a certain course of research, but there are other problems as well that one sees, particularly in the government. If you look at the structures of the Centers for Disease Control (CDC) and the National Institute of Health (NIH) for instance, NIH is divided into its institutes, CDC into its centers. Environmental health effects are in a center that's distinct from the Center for Infectious Diseases for instance. There is a question about where and how one would center a global warming activity in CDC, and I think there would be a parallel question at the NIH. Even the simple public health ramifications of what we're talking about here are too broad to fit any predefined specialty. What's required is a recognition that we're going to have to do an interdisciplinary effort and global climate will make strange bedfellows. Physicians and epidemiologists are going to have to get to know climatologists and atmospheric chemists and be able to deal with them on a similar set of terms and a similar set of frameworks. The global climate models or general circulation models will have to have a new input: the needs of the health researchers. We need to know, for instance, in health maybe more than in geophysical science, just what are the extreme situations. We need to know not just the mean, but what is the variability and, even more important than that, the sustained variability, because those are the situations in which health effects are particularly prominent. Something that Dr. Leaf mentioned, the idea of feeding back to our colleagues in other areas, is becoming extremely important. The fact that it's not occurring now is a problem.

DR. GIANNINI: I would like to limit my comments to infectious disease viewed as a potential problem with global warming and, more relevant to my own research, to ozone layer depletion. I think one of the problems is a matter of perception. First, in the United States physicians generally seem to feel that infectious diseases are no longer of major public health importance. This was not true in the past. The second problem is that issues of public health and preventive medicine get short shrift in medical education. For example, tropical medicine is not covered in most medical curricula. I think there are only perhaps four medical schools in the country that have an actual required course on parasitic diseases, so that our focus in the medical community is somewhat parochial. And third, I think the database for the possible health effects of UV on infectious diseases is really a relatively recent topic. Indeed, the immunological effects of UVB have only been elucidated within perhaps the last ten to fifteen years.

I have no doubt in my mind that, given the potent immunosuppressant effects of UVB, there will be an effect of UVB on the ability of the body to respond to infectious diseases. Any hard and fast information that we have to point to this as a possible health effect is missing at present and I think should be a topic for research.

DR. SHOPE: I agree with all of the above. I listed a number of infectious diseases that I thought might move into North America. This, in and of itself, was a parochial presentation. If we had included the world, I think we could have listed a lot more diseases. I agree that one of the reasons that scientists, and perhaps the public, haven't perceived health as a major consequence of global warming is that we don't have the information base on which to be sure. We don't have the information base on which not to be sure either. And I think that there are research needs, which I guess we'll come to to get this information.

DR. MACDONALD: Dr. Farrell, you started this discussion off with a consideration of public health. How would you respond to the comments that have been made?

DR. FARRELL: I buy everything except the point that you can't start until the regional estimates are better. There's a whole bunch of people who don't agree. For instance, all the secondary impact work has to be ongoing and parallel. If you wait until those regional estimates are really down to where you want them, it's going to be another fifteen years before you know what to do with them. The trick is to try to understand how to use the current estimates in your analysis, not that you're going to believe the data as it comes out, but it certainly will help you get a methodological development. Public health, along with agriculture, fisheries and forests, has to be considered in the development of the climate models, and we can't afford to wait until those models get done. Policy decisions will be put in place long before those models get the regional specificity that we feel comfortable with.

RALPH PERHAC (Electric Power Research Institute): I'm asking this question out of ignorance. Is it possible that we have not had sufficient emphasis on human health problems related to global climate because we may know what the solutions are to the problems that could exist? For example, if we're concerned about cataracts, we wear sunglasses and a hat. If we're concerned about skin cancer, we wear long sleeve shirts. Things of this sort. Maybe the real question is, not what to do about the health effect, but what are the conditions that would bring about a change in the health situation. In order to answer that, we're looking for the same sort of information that would apply to any other impact from global climate. That's where we may need the research. Is this why there's been less emphasis on health?

DR. KILBOURNE: Yes, it's difficult to predict health effects, and that leaves the possibility that the health effects may not be so severe or that they may be relatively easy to control. I could buy that hypothesis if, in fact, there had been a good and systematic study or series of studies that addressed the issue well and got us on target with respect to a plan. But what you're really seeing is a public health community and health community at large that doesn't have a plan, that doesn't have the faintest clue. The evaluations really haven't been done yet. I think that what you said is jumping the gun a bit and I don't think anybody could responsibly subscribe to that point of view right at this point in time.

DR. LONGSTRETH: I have to agree with Dr. Kilbourne. You picked an example, that of skin cancer and UV, where we do know better what the agent is and what the response is. That still doesn't help us get people to take that response. That's another issue entirely. When you get into the global change area where you're not looking at an agent like sunlight or UV, but at secondary and tertiary effects, then we don't have a handle on that. I've been trying to do some work on the whole relationship of health effects to air pollution but our database on air pollution, city by city, is very poor. We don't have a clue in many cases as to what the impacts are going to be without global change. If I were going to say anything, I'd say part of our problem is the fact that we don't have a good enough handle on what's going on right now.

BEVERLY TALTON (U.S. E.P.A.): I was the project officer on the tropospheric ozone criteria document. I want to share the perspective from that experience, from the National Ambient Air Quality Standards. Possibly you are facing public perception of acute versus chronic or distant effects. If we say that tropospheric pollution causes an incremental injury to the lung, we know that other stimuli do the same thing and maybe overwhelm the response compared with the tropospheric pollutants. For example, if smoking is deleterious to the lung, is ozone anywhere near as deleterious as smoking? So the public perceives that, "Oh well, it's just another added risk." If in fact you're talking about the possible development of fibrosis thirty years later in your life, the public pushes that aside even more. So it would seem to me that, if you're truly interested in getting the public aware of the public health implications of global climate change, perhaps a better approach would be to point out the emergency situations that can arise in public health by a breakdown of supplies of sanitary services, food supplies, the sheer loss of life from flooding and so forth. Cataracts are caused by other things, possibly. In the public's perception this is just another small incremental risk. Until you can focus on the acute emergencies that can befall the community or a city, I'm not at all sure that we're going to get their attention.

DR. FARRELL: I agree. I think that's precisely the problem. You could use analogous situations in other resource areas. If we're talking about reducing forests 50 years from now, all the Senators snore over it. If you say that you know the Southeast is going to have an economic problem because tree plantations are going downhill within eight years, and you're talking to a southern Democrat, it gets his attention. So I couldn't agree more with your comments.

AHMED MEER (Senior Science Advisor to the Assistant Secretary in the State Department): I noticed that, in the work on the Intergovernmental Panel on Climate Change (IPCC), public health has not been made a significant issue. I think it's worthwhile that you raise it here. I was, five years ago, U.S. Science Counsel in New Delhi and I also found it very difficult to get developing countries seriously concerned about the global warming issue and ozone depletion. I think it is fairly important that, if these health and agriculture effects of UV are identified, they would be helpful in getting support. But also it would be very difficult for the poor countries to handle the public health effects of global warming. I wanted to thank you for raising this issue and bringing it notice.

DR. MACDONALD: Thank you very much. I take it that there are a variety of underlying reasons why public health considerations haven't received attention, but I also gather that, in the future, we hope that the attention will increase and will be reflected, hopefully, in additional support for research.

DR. LONGSTRETH: I would like to relate that Suzanne Giannini, Hugh Taylor and I were sitting around a table last night commenting on the number of times we have gone to meetings and put forth an agenda about what should be done. I counted four, Hugh had six, Suzanne must have had about five and we're going to another one not too far away. I have to confess to a fair amount of frustration because you keep saying the same thing over and over again and we don't see any results from it. I'd be very happy to hear a way of making what we come out with more effective, which would somehow galvanize the money grantors to listen to us. Maybe it's a risk communication problem. Maybe we're not getting across the message.

DR. GIANNINI: I think there are more and more conferences to go to, to the issue of how to spend less and less money but we need to find a way to make more money available to look at these issues.

HERMAN F. COLE, JR. (Adirondack Park Agency): I'd like to underscore what Dr. Perhac was saying. I'm a layman, I'm not a scientist. I'm also a clergyman, which doesn't mean a great deal to many people except if they're interested in moral questions. There seems to be a great deal of disparity between the necessity for scientific research to be as specific and detailed as possible and our awareness of this problem as being global. Still the best way for me, as a layman, to get a handle on the entire problem is to try to think a bit, à la Paul Erlich, about things like mutualism as an essential component for this planet and its diverse life forms and the need to interact with each other and to support each other. I'm not going into the Gaia hypothesis kind of thing. Perhaps we should think in terms of Lynn Margolis's microcosmos and her attempts to show the history of how microorganisms help to develop the very organs of our own physical being, if you can buy that kind of thing. And what one sees as he looks out upon the problem we're faced with, global climate change, whether it be UVB or the raising of temperature, is that somehow or other there is a disruption occurring to the whole ecological system of this planet. In so far as one perceives the problem in that kind of dimension, it's extremely difficult to move down to specifically looking only at infectious diseases and looking only at specific kinds of impacts which we can dismiss because they are problems that have been dealt with in the past and can be dealt with in the future. In other words, there seems to be an inability to think both cosmically and in the particular terms that science demands.

DR. KILBOURNE: I'm not totally sure I agree but it's such a productive and nice thought that I'm not sure that I want to disagree either. I think that global problems can have their solutions in specific thoughts and specific studies. Perhaps they really can't be translated.

DR. MACDONALD: I think we might now turn to what is to be the main emphasis of this panel. We'll be coming back to the issue of how we bring to the policy makers these very urgent needs for additional research in a number of areas and the bringing of the public health issue further out into the open. What we intend to do is to go around, one by one, and discuss what the research needs are as viewed from the particular perspective of the individual and hopefully to get responses from the audience and from other panel members.

I'd like to start off since I had the pleasure of opening this conference, and I do it in part to emphasize the uncertainties that do exist and that need to be resolved. Also I raise a very important question which gets to a theme that I heard from several of our panel members, that is the need to be much more sure of what is going to happen before we're willing to commit major research efforts. And "much more sure" means to me, in part, what will the climate be like in some geographical area over some time, where that geographical area is still undefined?

Number one on my list is actually not research. It's the very great need to increase the pool of people working on global change. One of the most productive things government could now do is establish a program of support for graduate students and postgraduate students in this general area. Dr. Farrell made reference to 5,000 in his overall area, and about 140 countries. I suspect many of those are not really full-time participants in the research effort. The number of people who really devote full-time to this subject is minuscule compared to the challenge of the scientific problem and the potential long-range social, political and economic impacts. That would be, by far, my greatest priority. More people right across the board, more in atmospheric science, more in the ocean sciences, more certainly in the public health arena, the medical arena, and the biological sciences. We need to get more people into this field at an early age and not depend, as some would have us, on retraining aerospace engineers who have devoted their life to other pursuits and now are to become, overnight, experts in this field. I think it is a field that requires great depth and great talent in the physical, biological and social sciences.

My next priority addresses precisely the question of what can you do about predicting climate. And here I'll have to get slightly technical in raising the following question which is central. Is it mathematically possible to model climate? Can you, from the underlying equations, predict what climate will be at a given time in the future in some particular location? There's a body of theory which goes under the name of chaos theory or nonlinear dynamics that says the answer to that question, very probably, is no. Basically if you look at the computer models, what do you have? You have a system of approximately 100,000 coupled nonlinear equations. We know that, in simple systems of three coupled nonlinear equations, you evolve in time in such a way that you can never make sure, if you start in a certain place, where you'll end up in a predicted place. It's a central question. What is possible in prediction? We know that it is impossible to make weather predictions for any significant time ahead. Maybe four or five days, because slight uncertainties in your initial conditions, that is in what the temperature or pressure is at various points, lead to very divergent final states. This is then an underlying question that needs serious attention from the mathematical community. How far can we go in predicting climate in the sense that we've been talking about? I feel confident that we can make estimates of what the average temperature will be, given the atmospheric composition. We don't need large computers to do that.

Next, how do we deal with the notion of extreme events? Averages don't kill anybody. It's the variance that does it and so how do we develop ways of analyzing what will be the frequency, intensity and duration of extreme events? Extreme events mean storms, storm surges, hurricanes or prolonged periods of hot or cold weather. How do you approach that? This is almost at the conceptual stage. It isn't quite, because there is a developed theory for extreme events but it has never been applied in a satisfactory way to the kinds of issues that we're talking about.

If we turn to the kinds of observations that we need to make, one critical class of observations deals with understanding how energy is trapped within the atmosphere. It comes down to the role of clouds largely, but also the greenhouse gases and their interactions. We have gross ways of describing it; we have now begun to analyze a wealth of information from satellites, the Earth Radiation Budget Experiment (ERBE) and others. But we have almost nothing in the way of looking up, seeing what the cloud is, determining what its properties are, and translating that into an understanding of how much heat is actually trapped and how this, in turn, influences the state of the atmosphere.

We have beautiful records of the variability of carbon dioxide, produced not only at Moana Loa but also at the South Pole and many stations from the South Pole to Alert in Northern Canada. We have no comparable data on the other greenhouse gases. We have spot measurements of methane, carbon monoxide, nitrous oxide and the chlorofluorocarbons. These are so stable that only a few measurements or a few stations are required to establish trends in geographical distribution. We have no program any place in the world to monitor oxygen, yet every time you form a carbon dioxide, you take two oxygen atoms out of the atmosphere. If we had that data, and if it were of the quality of the CO_2 data, one could make a much better estimate of the relative importance of the sinks for carbon dioxide in the oceans and the biosphere.

Methane is, in the end, going to be an extremely important constituent in all these arguments. We have already discussed that methane is not only a greenhouse gas, but is also important in reactions that take place in the stratosphere; therefore, it is very important in questions of eventual demise of the ozone layer. Yet we know very little about the sources of methane, particularly the northern sources, and we know even less about the methane hydrates as a potentially large source of methane.

Methane hydrates are the largest source of carbon that we know about other than the carbon in the oceans, and much of that carbon is in the inorganic form. A very important issue deals with the microbial action in soils, with how much CO_2 is produced in soils. Methane is produced in soils as a function of changing atmospheric conditions. One can make a long list, but I have highlighted these to emphasize that there are major uncertainties, but we have some idea of how to approach them. My guess is we'll never get all the data we need. And if that's the case, I would argue that we should get on with the job of understanding the public health consequences in the face of, not only an uncertain future, but an unknowable future.

DR. LONGSTRETH: That we can never get to regional predictions of climate change is probable, but I'm not sure I see the worth of collecting all this information on methane and CO_2 and so forth. I would support that endeavor if I thought it was going to give me better estimates of what is going to happen in the public health arena, but otherwise it's an intellectual exercise which I'm not sure I understand.

DR. MACDONALD: I'd argue it's anything but an intellectual exercise because the one thing you want to know is how much greenhouse gas is entering the atmosphere and how much is leaving, in order to get some idea of balance. Then you can make guesses as to what the composition of the atmosphere will be and you can predict. All you can say is that the conditions are going to get warmer and then you can use analogues with either past climates or current climates and try to argue from those rather than trying to make precise estimates.

DR. FARRELL: I don't think it's a question of not getting regional estimates; it's getting regional estimates in sufficiently narrow uncertainty ranges that become fruitful to use. If you look at the general circulation models (GCMs) over North America, the range in temperature is 12 degrees centigrade and the range in precipitation is 60 percent from a minus 20 to a plus 40. You can't do much with that. The uncertainty will never get down to the level that you would want, for instance, it's going to get 1.72 degrees warmer and we're going to have 6.7 percent more precipitation. We'll never get there but we're going to reduce that uncertainty.

The most disturbing thing in this whole field, though, may be the carbon cycle. Oak Ridge has been working on the carbon cycle since 1957. In the last ten years, we have reduced the uncertainty about key parameters to about 30 or 40 percent. At the same time, we've also introduced 50 percent more numbers that we now need to understand. It's not a question of the classical look at taking an uncertainty band and narrowing it down; it's more like a circle. You move the science forward and then you discover more things that you don't know. The whole process is going to be like that. You could be sitting up here with a bunch of foresters or agronomists and they'd all complain, "You've got to get me better estimates." Well, they're never going to get down to a desirable level. For instance, if it's a four-degree range, what you might want to do is step your analysis in half-degree intervals. You could come up with very interesting results. It could be that you don't see any effects until you get way out to the three- or four-degree changes. This is a common argument that runs through the impact arena.

DR. KILBOURNE: I'm just afraid of something that Dr. Leaf and I referred to: uncertainty as a negative incentive for pursuing this kind of research. Uncertainty is making us recommend against starting until such time as better estimates are available. I don't think I was saying that. We'll have to deal with uncertainty. I share some of what you were saying in terms of the problems of modeling, having presented a model yesterday. I would be the first to say that successful modeling of the public health possibilities in the face of extensive global climate change is doubtful. You're talking about modeling diseases which involve microorganisms which may or may not be simple; I don't think that they are. Attempts to model disease transmission have resulted in such complexities that such models don't often reflect reality when put into practice.

Another problem is displaced persons. It's come up a couple of times. We haven't really addressed it because it's one of the hardest things to address. It depends critically, not only on the climate events themselves, but on politics. If there is less food, is there, therefore, starvation in one particular part of the globe? What is the political reaction to that? What is the human element there? What do people in one country do to help those in another? Or, if they do nothing, is there then a war? Is there then political instability?

And, finally, there is an additional complicating factor. Discussions, such as the one we're now having, can ultimately impact on the political process and would serve as an input into any model that you would care to make of the ultimate manifestations of this at a human level. That makes the observer part of the situation and complicates this beyond comprehension. Certain things about microorganisms and vectors may or may not be predictable or modelable but, when you get down to things that involve behavioral adaptations and people, I would throw in the towel.

DR. SHOPE: I don't share the pessimism. I think we collect data even if we're not sure that we're going to be able to predict on the basis of them. Throughout history, we've been collecting data and we would have nothing if we hadn't done so. We have to collect data. As far as the health and the infectious diseases are concerned, I would agree that we don't have the data to validate the models. We don't know whether they're good or bad but we need to start right now to collect those data. We have identified at this conference several disease states and some infectious diseases that might change their territory or might move. We should model them, knowing that they're imperfect models and then, as we have more data, improve the models.

DR. KILBOURNE: I wasn't saying you shouldn't do that. I was saying that, if I'm asked to try to model 21st century political behavior in the face of an impending disaster, I don't think I could do that very well.

DR. MACDONALD: Dr. Farrell says general circulation models are going to improve if you cut down the resolution. I have no confidence in that. I think that the underlying mathematics is such that you may not be able to get an answer. But I pose it as a research question.

DR. PERHAC: One really valuable purpose for collecting data which hasn't been mentioned is to see if changes are, in fact, occurring or if whatever action we're doing exacerbates this situation or actually ameliorates it. It's critical that we keep finding out what's happening in the atmosphere even if we can't use that information to improve our predictive capability. It at least tells us what changes are occurring or to what extent the earth has resilience and can react to whatever actions mankind today is taking.

DR. MACDONALD: Well, that certainly is a very valuable point, just one of the current puzzles that I mentioned yesterday. We have these great fluctuations in input of CO_2 from fossil fuel into the atmosphere, yet these fluctuations are not reflected in the trend of atmospheric concentration of CO_2. It's a puzzle and, until we get a much better handle on it, we can't say that we really have as good a grip on that part of the carbon cycle as we may have assumed earlier.

DR. LONGSTRETH: First, I'd like to agree with some of the comments that Mike Farrell made earlier about the data needs. We almost lost the Landsat database in its entirety along the way. There's a similar situation in the health research field and it has to do with the Robertson Berger meter. This is the only source of ground-based UV measurements that has a continuous history over a fairly long period of time, and it's been nip and tuck to keep the thing going even at the modest cost compared to most scientific endeavors.

What we really need for the ozone depletion area is better data on what we're actually getting on the ground. There's been a recent report from Joe Scotto and his colleagues at the National Cancer Institute (NCI). If you look at the Robertson Berger meter, it does not detect a change in ground-level UV. The explanation I've heard for this is that these RB meters are located at airports, polluted areas, and the tropospheric ozone contributes a fairly powerful absorption property to the process. But, if we let those meters go, we're never going to have comparative data. I agree with you entirely, Dr. Perhac, that we have to have these measurements so that we know what we're doing to the system. The reason the RB meters are being dropped is that they're not the best thing around. There are better meters coming along that are going to be spectro-radiometers, able to measure the energy in a one-nanometer wave band. But we can't dump that historical data. We need to have those things operating side by side for two or three years so we can then go back and try to figure out what's been going on histori-cally using the input of the spectro-radiometers.

I would put monitoring information as a key priority in what needs to happen. It's been distressing to watch what's going on and see how, if things get cut, monitoring is first to go. It's not glitzy enough and so that's basically what gets cut. Without good monitoring data, you can't validate your models and you can't test hypotheses. You real-ly need those data.

As for projects, I have a list as long as your arm, but I think the first item is im-munosuppressive effects of UV and what they could do to infectious diseases. We don't have enough information and it has the likelihood of a very high impact.

The impacts on populations from ozone depletion, if they really translate into UV, are of primary concern. What goes on in the troposphere? How much does that com-pensate for what we're doing to the stratosphere? That's an open question too. But, if we're really talking about significance, what is a 10- or 15-percent increase in UV on the surface going to mean? In terms of what Hugh Taylor was saying, that probably is going to pass the threshold for getting photokeratitis on the beach. Photokeratitis is not a nice thing; if you've ever had it, it's extraordinarily painful and you never want to have it again. We have to gather the information and start looking at some of these issues.

When you get into the issue of global warming and the tertiary effects or secondary effects, you're not looking at the direct impact of an agent on the host. You have to as-sess what it does to a vector, the agent and the host, and whether or not these things over-lap. The complexities become enormous. I agree that we have to play with models. We have to look at models and see if we can predict things. That's the only way we will refine the models and we're going to need better data.

In the exercise we did for our report to Congress, we had a malaria model which had been translated from Kenya to the U.S. and we tried to play with it. Well, it's got some big holes in it. There are some important elements to the model—the host, inter-mediate host and the distribution of habitats—which we had to assume were constant, even though we knew they probably would not be under climate change. You learn where the flaws in your fabric are by playing these games and testing these models and that's how you can refine them. You just can't wait until you get the best data. You're going to have to start playing these games now and, when you get data, then you'll be able to better figure out how to use it.

DR. MACDONALD: I think that we are clearly making a correlation of surface UVB with stratospheric ozone but, if you really are to get reliable information, you have to mount a campaign. You will want to make tropospheric measurements of ozone as well. The British Antarctic Survey has now made observations of ground UV and they find a decrease consistent with what you would expect. I'm very glad for that. If we couldn't get that one pinned down, I think we'd really be in bad shape. It's clearly essential that we keep these instruments going long enough so that, if better instrumentation becomes available, you can make comparisons and interpretations. If we hadn't had the Dobson meters functioning in the Antarctic, we would never have found the ozone hole.

DR. LONGSTRETH: John Frederick developed a NASA model to use NASA satellite data to try and estimate UV flux on the earth's surface. Hugh Pitcher at the EPA took this model and basically made it portable to a personal computer and then tried to validate it with the Robertson Berger meters, the data from Scotto and the people at Temple. He found that he could get reasonable validation if he was looking at clear-day information but, on cloudy days, where aerosols become important, the model rapidly degraded. The nice thing about the model was that you were not restricted to the 30 or so locations where you have Robertson Berger meters. More effort needs to go into developing models using satellite data and then validating them with either Dobson meter information or RB meter information. At some point, you would not be limited to the number of sites on earth where you can put a monitoring station.

DR. GIANNINI: I'd like to reinforce what Dr. Longstreth said about the need for developing some good ground-level, ambient UVB spectra data, not only for the reasons that she stated, but also in terms of modeling infectious diseases or any effect of UV on the immune system or any other parameter. We really need to know what's out there in the environment, so that we can shape our laboratory UV light sources to reflect what may actually be going on at ground level. With respect to all of the studies that have been done on the effects of UVB on infectious diseases, the majority of them have been done using fluorescent sun lamps which emit a fairly large UVC component. UVC is not present, at least at the moment, at ground level. That would provide valuable information for refining models, infectious disease and other models, for effects of UV.

DR. LEAF: I had a chat with Dr. MacDonald and I want to assure you that he's really not as nihilistic about the hopes for learning something and coming up with good, wise public policy as maybe he sounded from his earlier comment. But before we're going to attract young people into careers in this area, they have to know that there's a certain amount of probability that what they're going to work on is going to be important and that the consequences of their research is going to bear some fruit that will be useful to society. There may be some uncertainties about what the climate will be in small areas and so forth, but it seems to me that anyone knowing that we're putting 5.2 billion tons of CO_2 into the atmosphere each year and knowing something about its properties of energy absorption and the emission of the other greenhouse gases will realize that there's a problem. They may not be able to come up with absolute certainty in the kinds of conclusions that they draw, but they should be able to assign some probability band as to how likely these changes are. On the basis of that, the public and our government administrations, in their wisdom, can decide what is appropriate policy. I think Dr. Longstreth is complaining about something that everyone who's a pioneer in the field is going to run into. She's been working on public health issues as long or longer than almost anybody else and she's finding that the response of society, or even of the medical profession in the public health field, is slow. Everything moves at a glacial pace and, when it's a topic that you're particularly interested in, it goes even slower.

But there is one thing that I'm very concerned about for our future, relative to these things. We've been living in the last ten to fifteen years in a society in which our young people are being trained and brainwashed to have the expectation that the ultimate goal of their achievement should be their own individual gratification and that the entrepreneurial system and the marketplace should be the ultimate deciders of everything that goes on. I'm afraid that, at a time when the world's gotten smaller, we're going to have to look, not at what the immediate benefits are to our society or to the individual, but at what the needs are for the global survival of our offspring. We've got to begin to educate our young people and our old people to the fact that there's more to our purpose on earth than to consume the earth's resources at an accelerated rate. Because it's only when we can begin to work together and look at what the future is and make plans that are going to safeguard this planet, that we'll be achieving what those of us here are interested in. And certainly that's where the public health issues are tied in.

Having made those rather obvious comments, let me make a few points. The first one is the problem with the population growth. That's the root basis for the problems. With population growing and more people expecting to live the lifestyle that they see in Hollywood movies, the earth's resources are going to be consumed at an accelerated rate. That's not an issue that we've talked much about here. There are research issues in family planning, birth control measures and so forth. There's a very large informational problem here and we've got to somehow reduce the amount of bigotry and intolerance that has come into discussions of this issue. And I'm not going to go into that, but I think the problems are self-evident.

Secondly, there may be effects of the various environmental changes on nutrition. If we're going to know what the temperature rise is going to do to the crop lands, we have to get some estimates that are reliable enough, not only as to what the mean temperature changes are, but what the variability will be, the sort of thing that Dr. Kilbourne was talking about in relation to effects on humans. We need to know the effects of the extremes on crop lands and on agriculture. Much of this is researchable. It can be done in small greenhouses where one can regulate the conditions of growth and actually see what the temperature changes will be. We need to know better whether there will be changes in distribution of rainfall. Again, I think people agree that there will be increases because of the warmer climate. But it does make a difference whether central portions of the continents are going to be a little dryer or whether monsoons are going to change, because it's on those specific kinds of information that the probabilities of agricultural changes can be stated.

One rather vague issue is that, if the temperature increases, there will be less snow on the elevated portions of the earth and, with that, there will be less of a reservoir of melting snow to provide water for the water tables that are important for agriculture later in the summer periods. We need to have some assessment of what those temperature changes will be in areas of high altitude.

We've heard rather conflicting statements as to what happens after the forests are chopped down. There's enough forests that have already been reduced that it should be possible to make some careful observations as to what happens to secondary growth and how much erosion there is from rain and wind. I'm sure it is available. Dr. Farrell said that we don't even know what the questions are. Even when we know what the questions are in the health field, we don't know where to go to get the answers. So we welcome the panels that we heard this morning that will help us get information we need.

The urbanization of crop lands – when I was recently crossing the Po Valley in northern Italy, I was shocked that I didn't see any agriculture. All I saw were new industries that had come up along the highway. It's hard for me to believe that, as these kinds of changes are occurring in some of the most fertile crop lands in the world, we're going to continue to have enough good lands to produce the increased amounts of food that the world will need. We need some assessments of how much of that land is going to cities, how much under concrete and so forth. These are determinable variables which we should be able to understand.

The rise in the sea level – we've heard the assessment as to how much land will disappear for a half-meter, one-meter or two-meter rise. Our speaker last night, Sir Crispin Tickell, expects that there may be thinning of the ice levels at the polar caps and he discussed the effects that this would have. So again, we should be able to come up with some limits that will allow us to make probability estimates.

The atmospheric pollution problem – how much is it damaging crops? It is being studied at the present time, such as the effect of the increase in CO_2 as a substrate for growth and whether that will stimulate agricultural yields more than it will damage them. Some of the literature shows that weeds may grow much better than some of the cultivated agricultural plants when you raise the CO_2. As these issues are researched, we'll have some better answers.

PERRY BERGMAN (U.S. Department of Energy): I'm going to address what a lot of people are very frustrated about: that no one is paying any attention to what they're saying. Really the crux of the problem is that there are a lot of people in the technical and scientific community who don't believe in this stuff. It's really that simple. And I have to deal with these people every day. I would say probably the only thing that's going to get people's attention is something of an extreme nature. For example, Dr. Wallace Broecker of Columbia has proposed the idea that, whatever the change, it might come in some kind of quantum jump, not in a linear way that a lot of us are talking about. If something like that happens, then it will be like the San Francisco earthquake. All of a sudden everyone was working on earthquakes. Yet, if you looked in the engineering literature, an article in the ASME mechanical engineering journal this month predicted that this country is going to have a major earthquake and that we're totally unprepared for it. That's probably the way the thing is going to play out.

DR. MACDONALD: Let me make the following comment. The general estimate of a few degrees rise for the doubling of CO_2 or the equivalent is a number that's held up for a long time. The first published estimate was by a Swedish chemist, Arrhenius, in 1898. His number was slightly higher, by five degrees centigrade. I remember the first time I testified before the Senate Interior Committee, in 1965, I used three degrees. Nothing much has changed. The basic physics are well understood. It was underlined by the National Academy Report in 1979 which said that we have searched and searched for some way of getting around warming, some mechanism that will either stop or reverse the process, and we've been unsuccessful. In 1987, a group of scientists from a number of countries said basically the same thing, that the greenhouse effect is real. We understand the fundamentals of it. The degree of warming and the timing of warming remains uncertain. That uncertainty persists today.

I know of no paper published in the refereed literature that questions the validity of the concept of the greenhouse and rejects the idea that you can expect warming. It is a straightforward consequence of the quantum mechanical properties of molecules and the second law of thermodynamics. Warming is what you're going to achieve; how much and when remain points of debate. There's a lot of controversy in the press but, when you really get down to what I call good solid science that goes through the normal process of being looked at and reviewed, there is nowhere near the range from "the greenhouse is total nonsense," that one sometimes reads about in the *Wall Street Journal*, to the other point of view that we're approaching disaster very rapidly.

DONNA ORTI (Agency for Toxic Substances and Disease Registry): You might say this is a moment of confession for me because, in preparing for this meeting, I started researching my own personal collection of references on the greenhouse effect. And I think, in regard to public opinion, there is just so much demand for people's time that the greenhouse effect is not an issue. That's what we face. But, as I started researching in preparing for this meeting, started looking at the literature that I had in hand, I became aware of the seriousness of this issue, but I don't have a suggestion on how you can reach the general public. The issues at hand are poverty, drugs on the street, paying the mortgage — there are just too many things for a citizen to be concerned with. This is something that's going to happen after they pass away and let's get on with the business of living.

DR. MACDONALD: I brought my stack of public opinion polls. Two in-depth polls, one of 1200 people and one of 1500 people, found that something like 60 or 65 percent of the population claimed they understood what the greenhouse effect was about. I found that to be a remarkable number. When people were asked if this is a serious problem and how they would rank it, the greenhouse was not considered an outstanding problem, but what is happening to air was. That was the second issue of concern in both polls behind crime and drugs. It was the general issue of what's happening to the atmosphere of which the greenhouse was one component, urban smog being another, along with acid rain and all the other things we know about. There is a developing body of public opinion that is aware of it. It's certainly a subject that is very much on the minds of the Congressional part of our government. Just witness the number of pieces of proposed legislation that have been introduced.

MICHAEL MCCAULLY (University of Chicago): I wasn't going to offer a question so much as an observation, then maybe a question. The observation is that we tend to look at public opinion polls and say that's the public. We all know that there is one component out there who would say that there's no problem and who wants to get on with living. There's a third component of, I guess I would say, the activists. There are numerically lots of American citizens, particularly the well-to-do, who read and are socially concerned, who have viewed the phenomenon that we're talking about, taken the concern seriously to heart and feel that something has to change. There is a sense of finite resources on a global spaceship and a need to do something about it. And I wonder if we're not missing a connection between the scientific community contained in this room, persons working with the issue, and another group that we shun a little bit because they tend to exaggerate the information. They tend to think of gloomy outcomes, but they are concerned. On April 20th, Earth Day, there's going to be a major national program and I suspect it will get lots of press coverage and will reflect concerns about a lot of the issues that we're raising. But we're all going to wrinkle our brows a little bit because there's going to be a lot of deviance from the science as we would look at it. It reflects a very profound concern amongst a significant part of the population that there's something going on that needs to be acted on now. Is there some way that we can connect what we're talking about and doing to that community?

DR. MACDONALD: It's a very important question. Obviously organizations such as the one you're associated with, Physicians for Social Responsibility, can play a very important role. But I'd like to come back to that after going through our research needs.

DR. GIANNINI: I'd like to ask Donna Orti a question. Perhaps the problem that the public has with global warming as being very important is a matter of semantics. If I didn't know what the ramifications were, I might think global warming is not so bad. What if we gave it the name "earth over-heating"? Doesn't that sound more dramatic? Do you think that would raise the level of awareness?

MS. ORTI: I can't comment on the semantics, but I would fear that we would begin to sound as if we were overreacting to the situation. There's been several books published that seem to bring the worst out of a situation without considering, for example, that if we do have global warming, it's probably not as serious as a nuclear holocaust, at least not in the foreseeable future. But it's something that needs to be acted on. Educating the public on the seriousness is something that needs to be handled aggressively, but certainly with a certain amount of care. If treated in an extremist fashion, it would lose its validity. I was speaking to you as a member of the public and how I initially reacted to these issues or even in regard to the ozone thinning. Now we've had something dramatic happen. We've had an ozone hole. To me that's frightening. That frightens me more than global warming. And there's nothing we can really do to repair that. The CFCs are in the air and they're still collecting. It's going to probably worsen. But how do you tell people who don't have a science background about the significance of this without truly alarming them? We're taking steps to resolve the CFC issue, but the damage has been done. The horse is out of the barn. In regard to global warming, I think we could use the CFC issue as something that precedes, or as an example of what could happen to, global warming if we don't take steps now.

One other thing I'd like to mention is, even if we don't have global warming, if it doesn't go up two or five degrees, we as an industrialized nation need to be concerned about recycling, depletion of natural resources, deforestation. These are all issues that I think can be taught in our schools, without even talking about the ramifications of global warming. If those things are adopted as a national policy or world-wide policy, global warming may in fact become a non-issue.

DR. FARRELL: To quote a friend of mine, this might be an "infrared herring." We have 130 bills in front of Congress. We probably had, in the last 18 months, over 200 global change meetings. We have nations who have adopted global change related policies already. Australia has already put a one-meter rise into its sea level ordinances and its building codes of the coast. You have a program that's gone from 100 million dollars to 200 million dollars and which admittedly is probably still underfunded. Frankly I don't need much more public awareness of this issue. I think we have it. We have the IPCC, which is a horrendous undertaking, never before seen in the history of the environmental movement. As a scientist working in the field, I don't see this lack of public awareness. What I do see is a public that is confused. There is confusion among ozone, the greenhouse and CFCs. Somehow they mashed all those things together. I don't see this lack of public opinion. What I see is a very poor educational system. There's no curriculum for this in the schools. If you look at what's available to the high school science teacher, it's pretty mediocre. The only way you're going to get that into the high school curriculum is by giving them a prearranged and orderly package they can use. They just don't have the time to develop those things themselves.

STATEMENT FROM THE FLOOR: I've noticed a small minority opinion or a school of thought developing that claims, if you consider the five billion people on earth right now, a rise of temperature of two or three degrees would be beneficial for the greater majority of the population. I'm not saying it would or wouldn't; I'm just saying this is a school of thought that is developing. Here in America, the problem may not be global climate; it may be resistance to change. Maybe it's not good for us but, for the rest of the world, perhaps it may be a benefit. If that's so, then I think we have to start examining our own thinking very clearly.

DR. MACDONALD: I quite agree with you that there is that school of thought. A strong proponent of that view is a leading Soviet climatologist and, coming from Leningrad, he may have a point.

On the other hand, I come back to the fact that average temperature or temperature changes are in a sense a very poor measure. Really what counts are the extreme events: frequency, intensity of storms, heat waves, and so on. I'm not certain that in a warmer world hurricanes are going to be more frequent and intense. Preliminary arguments suggest that that's the direction we're going to go. In which case, I can't see that any place in the world is better off. But maybe that's even too simplistic a view.

STATEMENT FROM THE FLOOR: The United States science leads the world in this issue and heads many of the international meetings and committees. We are listened to and we do have an obligation to do something about this. I find it very parochial, when I'm back in the United States, that we seem to not recognize this as a global problem. We cannot solve this problem without international cooperation. We have, for example, tremendous knowledge and know-how in this country. We've got to share our knowledge and know-how whether it's for CFC technology or for other areas. Finally, there's a large part of the world that does not like sharing data, something that we have done well. Our scientists and our way of doing science are respected around the world. If we promote that, we will solve this problem in a better way than we are today.

STATEMENT FROM THE FLOOR: A paper just came out recently from the holder of an endowed chair in climatology at MIT and he questions very seriously whether we're able to judge that a two- or three-degree rise in temperature would necessarily bring on extreme events. Now this is not a nut who says this.

Secondly, with regard to temperature rise, when I said that the school of thought is developing which claims that a two- or three-degree rise in temperature could be beneficial for more people than it would be adverse, these people are not considering just temperature rise, they're looking at other things that go with it. For example, there's a good possibility that a rise could increase precipitation and, for many parts of the world, that could certainly be a positive benefit. We'd better be careful that we don't become too provincial in our own thinking about what may happen just to the United States. There's no guarantee that a temperature rise would necessarily be bad for all of the United States. For example, many of the models show that there would be an increase in soil moisture in the midwest. We always talk about heating up, but the models will show you that, in some parts of the United States, the overall rise in temperature could result in cooler local temperatures and increased soil moisture even in the farm belt. These are far from resolved issues.

DR. FARRELL: In a recent report that's just been completed, they looked at the precise issues that you're talking about in terms of long term and short term. In the short term, there will definitely be winners and losers. There will definitely be some short-term advantageous positions to be in, but this study concluded that in the long term it's a lose/lose situation for everybody. So add that to the fuel. I don't know who you want to believe. At least there's a separate opinion.

DR. KILBOURNE: I think we risk sounding like a broken record. The typical scientific paper or presentation ends up with the phrase, "Further research is needed." That's because often, in a scientific study, you open up as many questions as you are able to answer. Naturally one involved in a project finds that, "Yes, now we've got to do this or that." There are limited amounts of resources for doing research and you have to make priorities. I would argue that this is a priority area because of the economic and political ramifications that are so enormous, and the health consequences, while we admittedly know little about them, are potentially grave. Not knowing is reason enough for action and that would be the basis on which I'd put forward the need to pursue further research.

I see four major areas in which global warming might have an important impact and I'd like to enumerate them and discuss all but one briefly.

The first is illness related directly to the heat. There is actually a fair amount of basic literature in this area, principally from physiologists. There is a very embryonic epidemiologic literature that will be of much use ultimately. There are real world, here-and-now reasons to pursue a research agenda. The fact that we have cities in our country which, on a reasonably regular basis, exceed temperatures wherein there is significant morbidity and mortality due to the heat argues that we need to further quantify the times when those events are going to occur and set up models that are likely to be predictive of those events. We find our ability to put prevention systems into place and look at personal risk factors will allow us to identify the people at whom such intervention should be directed.

Secondly, temperature has an effect on chronic diseases and other conditions. This is not something that we've addressed very much in this conference. When you count the number of deaths that physicians attribute directly to the heat during heat waves and compare that with the apparent mortality excess, you find that physicians recognize only from 10 to 50 percent of those deaths as having been due to the heat. What you're apparently seeing is a rise in deaths from other types of disease. Largely, in past heat waves, those have been cardiovascular disease and cerebral vascular disease. I used to not believe that. Recently there is some physiologic literature which appears to support the idea that blood becomes slightly hypercoagulable when a person is subjected to heat stress. So in a partially occluded artery in the heart with atherosclerosis, the finishing touch, usually a thrombus that forms in that area, could occur under the stress of an extreme event. It is not so unlikely at all. That also occurs in the brain and causes stroke.

There are other things that have only recently come onto the public health agenda which are not precisely chronic diseases. I'm thinking of things like homicide. There have been a couple of reports of heat waves in which death at the hands of others has risen. Exactly what the mechanism is there is not understood. One gets into a lot of psychological and sociological underpinnings to the whole phenomenon of homicide. I don't think we should forget about it though.

Going back to the chronic diseases for a second. As the population gets older, the proportion of persons with diseases associated with aging increases – those chronic diseases which are possibly going to be affected by extreme events in a manner similar to or parallel to that which I just mentioned. All of those things can be correspondingly more important simply for demographic reasons.

Another big area is that of infectious agents. I'd rather not stick my neck out in the presence of persons more expert than I am, but I see both vector and non-vector borne diseases as potentially important.

And finally, the fourth area, which in some sense crosses along the bounds of the other three, is that of displaced persons. If we have substantial numbers of refugees as a result of extreme conditions, we are going to have problems that comprise all of the problems of that class of people. People will have to live under conditions that don't involve the usual sanitary precautions and interventions that are characteristic of modern developed cities. There will be problems potentially with airborne diseases spreading rather rapidly, with enteric diseases from the absence of sanitary conditions, and all of the diseases that correspond to less civilized situations and less civilized times may flourish.

That is a big area. It's the least subject to the modeling kind of approach and one of the ones in which judgment, politics and policy are going to be absolutely key. The politician's input may, in some sense, be more important than that of the scientist with respect to some of these possible outcomes, particularly that last one.

Finally, I would argue the need for this research even if we are not certain that things are going to get worse. I don't pretend to be that familiar with all the predictions as some other people are, but I think the potential consequences here are so bad that they should motivate us rather strongly.

DR. GIANNINI: I'm going to restrict my remarks to infectious diseases and I'd like to say that this may seem to be an esoteric subject for people living in the temperate zones but in fact even in the United States, until very recently, this was not the case. We had malaria in this country up until just after World War II and a colleague of mine at Columbia University, Dr. Roger Williams, who was very active in mosquito eradication programs, said that just before they successfully managed to eradicate the mosquitoes, we had three consecutive years of abnormally cool weather. I don't know if there may be a connection, but it certainly is thought provoking. In my own family, I have two aunts who died of diphtheria in childhood. And certainly, I'm sure that all of us have had relatives who've died of infectious diseases in our parents' generation or grandparents' generation. The way that we've managed to control infections in this country has largely been due to the development of drugs, the widespread use of vaccines, improved nutrition and improved hygiene. We've heard evidence that many of these preventative measures, particularly nutrition and hygiene, could potentially be compromised by global warming. If we limit our consideration to the effects of ultraviolet radiation on the immune system, there are two points to bear in mind. One is that chemotherapeutic agents work very poorly in immuno-compromised individuals. And certainly with respect to Leishmaniasis, we're seeing now in Europe people who are critically infected with Leishmania in childhood or in early adulthood and subsequently develop AIDS. They then come down with visceral leishmaniasis which is virtually refractory to drug treatment.

Toxoplasma gondii is a protozoa and parasite very prevalent in the population of the United States; about 10 to 20 percent of the general population is infected. Usually in an immunocompetent host, it doesn't cause any serious disease but, in an immuno-compromised person such as someone who is either immuno-suppressed for reasons of organ transplantation or who maybe has developed AIDS, this can develop into fatal infection.

Here we have the possibility to examine directly a definite effect of ozone layer depletion increases in ambient UV levels. Perhaps the most important priority in research would be to look at the ability of an individual who is exposed to various ambient doses of UV to be successfully vaccinated. If it turns out that infectious diseases will be a serious public health problem due to global warming or to any other kind of environmental influence, vaccination will be one of the weapons in our armory. It's not at all certain that an individual who is exposed to high levels of UV in the environment can respond to a vaccine to develop protective immunity. This question has not been addressed at all.

Another area that is important in this field is more accurate modeling of UV effects on infectious disease animal models. We've already mentioned the necessity to acquire some ground-based UVB data which can give us some idea of what wavelengths are out there and at what levels. We can then apply them in the laboratory to try to mimic the situation in nature. We need some information as to how close the immune response of the mouse is to the immune response of man in terms of effects of UVB. There's not a lot of data on this but what data there is indicates that, both in mice and in humans, there is a strong immunogenetic component to the ability to be immuno-suppressed locally by UV radiation. That ability is not linked at all to skin pigmentation. Low doses of UV radiation may suppress the local development of contact hypersensitivity and may inactivate Langerhans cells in the skin. Similar doses are as effective in humans as they are in animal models. This suggests that the mouse models may be valid.

Another area of research that is important is the long-term chronic effects of UVB on infectious disease. If we limit ourselves to the tropics and the semi-tropics, many bacterial and parasitical diseases are chronic and they persist for decades. An immuno-compromised individual may not be able to eliminate the infecting organisms from the body and, therefore, even though they may superficially look fine, they're in a carrier state and are able to seed the environment with additional infectious agents. People in highly endemic areas may be subjected to repeated infections with the same infectious agent over and over again so that effects on immunity are very important. These have been very little studied.

Almost all the studies that have been done have looked at acute exposure. We don't know whether individuals may become tolerant to the immunological effects of UV if they are at risk only in cases of acute exposure. Very little is known about the effects of chronic exposure to UV but there is some evidence in mice that chronic exposure leads to shortening of the life span. Chronically sun-damaged skin in humans seems to be immunologically compromised. Those areas on the body respond poorly to being sensitized to contact allergens or to injection of tuberculin or other common antigens. Those are the areas that I would target for priorities.

DR. SHOPE: I would like to add to what Ed Kilbourne said on infectious diseases and limit my remarks to infectious diseases that are bound to the ecology. We need research programs which would be multidisciplinary and aimed at some identifiable infectious diseases that might change in prevalence secondary to global warming. We can identify the diseases or at least some of the diseases. I would envisage the program as being based in universities primarily where you have multidisciplinary groups and probably the monies for this could be catalytic monies. In other words, set up the organization and the interaction of the various groups and then various research projects could be funded through some of our normal channels.

The places where I would look at ecologic changes would be on the fringes of the range of the vectors for vertebrate reservoirs of the identified diseases. That's the area where you're most likely to get information which might give you clues as to what will happen if temperature or rainfall or other climate conditions change.

The only other conditions that I can think of which might be worth studying would be what happens when man artificially intervenes, such as when we build dams and change the temperature and humidity in an area artificially. Some studies of this nature have been done but I think we need more. If I were going to pick areas that I thought were relevant to the United States — and I realize we're talking about global change — I would look at Mexico because that's where the tropically-based diseases are now that we might find. I would also look at Alaska, which is the territory closest to the extreme changes that are predicted, and study the ecology of some of the creatures that are in that area.

We also need ongoing surveillance of these diseases. Over the next ten or twenty years we need to know what is happening, which is really very different from the research aims. We should not forget the data gathering that we need to tell whether the changes that we're predicting, if and when they occur, actually have an effect or not.

DR. FARRELL: I would like to see balance in the national program. I would like to see everything of importance handled. The way I would do this, and I'm surprised that your community hasn't done it yet, is to do comparative risk analysis and try to show the position of public health in the whole scheme of things. Is public health more sensitive than fish, forestry and agriculture? Depending on how you frame that analysis, it might be quite enlightening, especially to Congressional people, getting their eye and attention.

The discussion on climate models is good as far as it goes. However, it's only half of the equation. How much temperature and precipitation are going to change in the extreme events is the half that that represents. The other half is when. It's the timing that's way off, by 300 percent. It's very critical that we understand just when CO_2 is going to double. That's less than clear now. The uncertainty in the timing is 80 years, a fairly large gap.

I'd like some more money spent funding research into the timing of the effect, namely, when will CO_2 double. I couldn't support more the argument suggested here on data.
That's going to be crucial to our understanding. It's been crucial in every other area, so
I can't see why public health would be any different. I think the studies in St. Louis
epitomize the very detailed work you're going to do.

This brings up a very critical problem that hasn't been brought up here and probably is unfamiliar to you. When you do these detailed problems, you have a problem of scaling up. We deal with it all the time as ecologists. We all are data site-specific and yet
we're asked to talk about North America. There's a lot of problems of taking this very
site-specific information and scaling it up to national boundaries or continental regions.

The last thing is the multidisciplinary approach. I think this is semantics. Multidisciplinary to me is when you get a bunch of good scientists all working independently on
the same project and there's very little cross-fertilization and very little communication.
I don't think that's what you mean. We use the term interdisciplinary. The public health
officials can't work in isolation from the climatologists. They are actually going to have
a difficult time working in isolation from groups like ours at the Information Center because we have some of the keys to the problem. I do not understand all these "miases"
and everything we heard about for the last three days; I'm sure that the same thing holds
for some of the problems in understanding the climate related parameters. So I'm very
encouraged at the end of this conference.

DR. MACDONALD: This afternoon you've heard a discussion of some of the research needs. There are some broad areas that require much attention, ranging from just
what we understand about climate and how well we can predict it to the details or the
kinds of questions that surround propagation of infectious diseases. The whole issue of
impact of both temperature and heightened UV radiation needs further work. What
comes out of this is that there is really a very rich agenda of things that, not only should
be done from a research point of view, but are critical in framing the longer-term policies.
To my mind, global change is not an issue that's going to be resolved this year or the next
or in the next decade. It's going to be with us for a very long period of time and policy
will evolve incrementally. It will be in spurts that will follow, I expect, the IPCC Report
of next year. On the whole, one can see that over the years, as more information comes
in, government policies, both here and in countries abroad, will evolve in response. We
need to recognize that some things may not be delineated within any finite time, and it's
in those areas where policy makers will have the most difficult task. If we're to assist the
policy makers, it is our responsibility to push for the kinds of research programs that
have been outlined today.

As a follow-up to this meeting, we need to consider in greater detail the message
that has come across exceedingly well at this meeting. We hope it can be communicated
effectively to those people who have responsibility for setting national research policy.

I would like to end by thanking the people, particularly those at the Center for Environmental Information, who have made this meeting possible. I also want to thank the

distinguished panelists for their incisive comments delivered with a great deal of feeling and, most of all, the audience. Thank you very much.

CONFERENCE PARTICIPANTS

Christopher J. Allabashi
Nixon, Hargrave, Devans and Doyle
Washington, DC

Thomas Anding
University of Minneapolis
Minneapolis, MN

Joseph Bangiolo
National Institute of Allergy and
Infectious Diseases
Bethesda, MD

Gerald Barton
National Oceanic and Atmospheric
Administration
Washington, DC

Perry Bergman
U.S. Department of Energy
Pittsburgh, PA

Patricia Bownes
Citizens for Alternatives to Chemical
Contamination
Highland Park, MI

Susan Brackett
Investor Responsibility Research
Center
Washington, DC

Andrea M. Brock
Legacy International
Alexandria, VA

Carl Casebolt
National Council of Churches
Washington, DC

Peter Clerry
Environmental Defense Fund
Washington, DC

Herman F. Cole, Jr.
Adirondack Park Agency
Ray Brook, NY

William Coleman
Electric Power Research Institute
Palo Alto, CA

Wendy Cortesi
National Geographic Society
Washington, DC

Debra Daigle
Unistar
Washington, DC

Raymond Daynes
University of Utah Medical Center
Salt Lake City, UT

Marta Dosa
Syracuse University School of
Information Studies
Syracuse, NY

Clay E. Easterly
Oak Ridge National Laboratory/Martin
Marietta Energy Systems, Inc.
Oak Ridge, TN

Karl Esch
Institute of Electrical and
Electronics Engineers
Washington, DC

James R. Fouts
National Institute of Environmental
Health Sciences
Research Triangle Park, NC

Mary Gant
National Institute of Environmental
Health Sciences
Research Triangle Park, NC

Lester D. Grant
U.S. Environmental Protection Agency
Research Triangle Park, NC

Nelson E. Hay
American Gas Association
Arlington, VA

Cheryl Hogue
World Climate Change Report
Washington, DC

Anna Inglis
American Gas Association
Arlington, VA

Joan M. Jordan
National Science Foundation, Division
of Atmospheric Sciences
Washington, DC

Laurence S. Kalkstein
U.S. Environmental Protection Agency
Washington, DC

Robert Engelman
Scripps Howard News Service
Washington, DC

Michael P. Farrell
Oak Ridge National Laboratory
Oak Ridge, TN

Chris Fuchs
Center for Global Change, University
of Maryland
College Park, MD

Susanne Holmes Giannini
University of Maryland School of
Medicine
Baltimore, MD

Keith Haglund
Medical Tribune
Washington, DC

Ann Hirschman
Concern, Inc.
Washington, DC

Ferdinand W. Hui
National Institutes of Health
Bethesda, MD

Ann Jones
Rochester Regional Group, Sierra Club
Rochester, NY

John R. Justus
Congressional Research Service,
Science Policy Division
Washington, DC

Edwin M. Kilbourne
Centers for Disease Control
Atlanta, GA

Donna L. Kraisinger
Amoco Corporation
Chicago, IL

Lisa D. Krueger
Illinois Power Company
Decatur, IL

Paul A. Locke
Environmental Law Institute
Washington, DC

Richard H. Lovely
Battelle Northwest
Seattle, WA

J.M. Masterton
Canadian Climate Centre/Environment
Canada
Unionville, Ontario, Can

Robert H. McFadden
Motor Vehicle Manufacturers
Association of the U.S.
Washington, DC

S. Ahmed Meer
U.S. Department of State
Washington, DC

Stanton S. Miller
American Chemical Society
Washington, DC

Katy Moran
Congressman Scheuer
Washington, DC

C.R. Krishna Murti
Madras Science Foundation
Adyar, Madras, India

Randi Kristensen
Physicians for Social Responsibility
Washington, DC

Alexander Leaf
Harvard Medical School
Boston, MA

Janice Longstreth
Clement Associates
Fairfax, VA

Gordon J. MacDonald
The MITRE Corporation
McLean, VA

Michael McCally, M.D.
University of Chicago
Chicago, IL

Blanche M. McIntyre
Southern Company Services, Inc.
Birmingham, AL

Douglas Mewett
Ministry of the Environment, Acid Rain
Office
Toronto, Ontario, Can

Linda Moodie
National Oceanic and Atmospheric
Administration
Washington, DC

Barbara Scott Murdock
Health & Environmental Digest (The
Freshwater Foundation)
Navarre, MN

Beth Nalter
Environmental and Energy Study
Institute
Washington, DC

Madeline Nawar
U.S. Environmental Protection Agency
Washington, DC

Mike Nazemi
South Coast Air Quality Management
District
El Monte, CA

Will Ollison
American Petroleum Institute
Washington, DC

Donna Orti
Agency for Toxic Substances and
Disease Registry
Atlanta, GA

Roland Paine
American Meteorological Society
Washington, DC

Ralph M. Perhac
Electric Power Research Institute
Palo Alto, CA

Susan Ann Perlin, Sc.D.
U.S. Environmental Protection Agency
Washington, DC

Roger Polisar
Albuquerque Environmental Health
Dept., Air Pollution Control Div.
Albuquerque, NM

Tom Prugh
National Environmental Development
Association
Washington, DC

Paul Recer
Associated Press
Washington, DC

John R. Robinson
Natural Resources Defense Council,
Inc.
New York, NY

Randy Schmidt
Associated Press
Washington, DC

T. Schneider
National Institute of Public Health
and Environmental Protection
3720 BA Bilthoven, Neth

George W. Sherwin
Sunsor, Inc.
Pittsburgh, PA

Robert E. Shope
Yale University School of Medicine
New Haven, CT

Jerry G. Simpson, M.D.
Marathon Oil Company
Findlay, OH

Linda Spencer
Infoterra, United Nations Environment
Programme
Washington, DC

Marnie Stetson
Environmental Law Institute
Washington, DC

Barbara Stocker
S & EH Information Center, Monsanto
Company
St. Louis, MO

Frederick W. Stoss
Center for Environmental Information,
Inc.
Rochester, NY

Barbara C. Tansill
American Chemical Society
Washington, DC

Elizabeth Thorndike
Center for Environmental Information,
Inc.
Rochester, NY

Beverly E. Tilton
U.S. Environmental Protection Agency
Research Triangle Park, NC

Robin Tuttle
Renault USA, Inc.
Washington, DC

Jaroslav J. Vostal
General Motors Corporation,
Environmental Activities Staff
Warren, MI

Linda L. Wall
Center for Environmental Information,
Inc.
Rochester, NY

Carol Werner
Environmental and Energy Study
Institute
Washington, DC

R.H. Wheater
American Medical Association
Chicago, IL

David Wooddell
National Geographic Society
Washington, DC

Janine Wright
American Public Health Association
Washington, DC

Douglas J. Yarrow
British Embassy
Washington, DC

Hugh Taylor
The Johns Hopkins University
Baltimore, MD

Crispin Tickell
United Kingdom Mission to the United
Nations
New York, NY

Jacqueline H. Trolley
Institute for Scientific Information
Philadelphia, PA

Peter Van Voris
Battelle Northwest
Washington, DC

William R. Wagner
Center for Environmental Information,
Inc.
Rochester, NY

Sterling L. Weaver
Nixon, Hargrave, Devans and Doyle
Washington, DC

Leif E. Westgaard, M.D.
Embassy of Norway
Washington, DC

James C. White
Center for Environmental Information,
Inc.
Rochester, NY

Geoff Worton
American Institute of Aeronautics and
Astronautics
New York, NY

Ronald E. Wyzga
Electric Power Research Institute
Palo Alto, CA

Marshall Yates
Public Utilities Fortnightly
Arlington, VA

CONFERENCE PROGRAM

Program

Tuesday, December 5, 1989

Public Health Impacts of Atmospheric Change

Moderator: **Dr. James R. Fouts**, Senior Scientific Advisor to the Director, National Institute of Environmental Health Sciences

Overview of Global Atmospheric Change
Dr. Gordon J. MacDonald, Vice President and Chief Scientist, The MITRE Corporation

Immune System and Ultraviolet Light
Dr. Raymond Daynes, Head, Division of Cell Biology and Immunology, University of Utah Medical Center

Effects of UVB on Infectious Diseases
Dr. Susanne Holmes Giannini, Center for Vaccine Development, University of Maryland School of Medicine

Infectious Diseases & Atmospheric Change
Dr. Robert E. Shope, Department of Epidemiology and Public Health, Yale University School of Medicine

Moderator: **Dr. Laurence Kalkstein**, U.S. Environmental Protection Agency

Human Nutrition and Atmospheric Change
Dr. Alexander Leaf, Chairman, Department of Preventive Medicine, Harvard Medical School

Cataracts and Ultraviolet Light
Dr. Hugh Taylor, Associate Director, Dana Center for Preventive Ophthalmology; Associate Professor of Ophthalmology, The Johns Hopkins University

Cancer and Ultraviolet Light
Dr. Janice Longstreth, Executive Director, Division of Science and Policy, Clement Associates

Respiratory Effects of Atmospheric Change
Dr. Lester D. Grant, Director, Environmental Criteria and Assessment Office, U.S. Environmental Protection Agency

Temperature Related Health Effects and Atmospheric Change
Dr. Edwin M. Kilbourne, Chief, Health Studies Branch, Centers for Disease Control

Dinner
Speaker: **Sir Crispin Tickell**, British Permanent Representative to the United Nations

Wednesday, December 6, 1989
Information Sources and Needs

Moderator: **Dr. Marta Dosa**, Professor, School of Information Studies, Syracuse University

Panel:

Joseph Bangiolo, Chief, Evaluation Sector, Information Technology Branch, National Institute of Allergy and Infectious Diseases

Gerald Barton, National Oceanic Data Center, National Oceanic and Atmospheric Administration

Donna Orti, Program Specialist, Health Education Program, Agency for Toxic Substances and Disease Registry

Linda Spencer, Information Specialist, Infoterra, United Nations Environment Programme

Geoff Worton, Director, Professional Services, American Institute of Aeronautics and Astronautics, Technical Information Service

Research Priorities

Moderator: **Dr. Gordon J. MacDonald**, Vice President and Chief Scientist, The MITRE Corporation

Information Needs and Research Priorities: Bridging the Gap
Dr. Michael P. Farrell, Director, Carbon Dioxide Information Analysis and Research Program, Oak Ridge National Laboratory

Panel:

Dr. Raymond Daynes	Dr. Alexander Leaf
Dr. Susanne Holmes Giannini	Dr. Janice Longstreth
Dr. Lester D. Grant	Dr. Gordon J. MacDonald
Dr. Edwin M. Kilbourne	Dr. Hugh Taylor